Statistical Methods in Practice

Statistical Methods in Practice: for Scientists and Technologists

Richard Boddy

Gordon Smith

Statistics for Industry, UK

A John Wiley and Sons, Ltd., Publication

This edition first published 2009

© 2009, John Wiley & Sons, Ltd

Registered office
John Wiley & Sons Ltd, The Atrium, Southern Gate, Chichester, West Sussex, PO19 8SQ, United Kingdom

For details of our global editorial offices, for customer services and for information about how to apply for permission to reuse the copyright material in this book please see our website at www.wiley.com.

The right of the author to be identified as the author of this work has been asserted in accordance with the Copyright, Designs and Patents Act 1988.

Library of Congress Cataloging-in-Publication Data:
Boddy, Richard, 1939-
    An introduction to statistics for scientists and technologists / Richard Boddy, Gordon Smith.
        p.   cm.
    Includes bibliographical references and index.
    ISBN 978-0-470-74664-6
  1. Science–Statistical methods. 2. Technology–Statistical
methods. I. Smith, Gordon (Gordon Laird) II. Title.
    Q180.55.S7B63 2009
    519.5–dc22

                                                                        2009025914

A catalogue record for this book is available from the British Library.

ISBN: 978-0-470-74664-6

Set in 10/12pt Times by Laserwords Private Limited, Chennai, India

Printed and bound in Great Britain by TJ International Ltd, Padstow, Cornwall.

Contents

Preface

This is a practical book on how to apply statistical methods successfully. We have deliberately kept formulae to a minimum to enable you to concentrate on how to use the methods and to understand what the methods are for. Each method is introduced and used in a real situation from industry or research.

The book has been developed from the courses run by Statistics for Industry Limited for over 30 years, during which time more than 10,000 scientists and technologists have gained the knowledge and confidence to apply statistics to their own data. We hope that you will benefit similarly from our book. Every method in the book has been applied successfully.

Each chapter starts with a situation obtained from our experience or from that of our fellow lecturers. It then introduces the statistical method for analysing the data, followed, where appropriate, by a discussion of the assumptions of the method.

The examples have been chosen from many industries – chemicals, plastics, oils, nuclear, food, drink, lighting, water and pharmaceuticals. We hope this indicates to you how widely statistics can be applied. It would be surprising if you could not successfully apply statistics to your work.

To get the best out of it we would suggest you do the following:

- Study each chapter.
- Work through the problems using the *Crunch** software (available from the books, website at http://www.wiley.com/go/boddy) where appropriate.
- Check the solutions to your problems – not just the statistics; ensure you agree with the inferences made from the analyses.
- Think of situations where the statistical methods could be applied to your work.
  However, you must be aware that analysing data for its own sake is of little value; any analysis must enable you to make better conclusions.

Some of the worked examples in the book differ marginally from those given by *Crunch* due to premature rounding. We have also not followed our own advice on rounding given in Chapter 17. We have kept more significant figures in the calculations so that the reader can follow and reproduce them where he/she considers it to be desirable.

Statistics for Industry Limited was founded by Richard Boddy in 1977. He was joined by Gordon Smith as a director in 1989. They have run a wide variety of courses worldwide, including Statistics for Analytical Chemists, Statistics for Microbiologists, Design of Experiments, Statistical Process Control, Statistics in Sensory Evaluation, and Multivariate Analysis. This book is based on material from their Introductory (Statistical Methods in Practice) course.

Our courses and course material have greatly benefited from the knowledge and experience of our lecturers: Derrick Chamberlain (ex ICI), Dave Hudson (ex Tioxide), Sandy MacRae (University of Birmingham), Martin Minett (MJM Consultants), Alan Moxon (ex Cadbury), Ian Peacock (ex ICI), Joyce Ryan (ex Colgate Palmolive), Colin Thorpe (ex GE), Malcolm Tillotson (ex Huddersfield Polytechnic), Stan Townson (ex ICI), Sam Turner (ex Pedigree Petfoods) and Bob Woodward (ex ICI). In particular, we would like to acknowledge John Henderson (ex Chemdal) who wrote the Excel-based software *Crunch* which is used throughout the book and Michelle Hughes who so painstakingly turned our notes into practical pages.

Dick Boddy
Gordon Smith
Email: s4i@aol.com
2009

1

Samples and populations

Introduction

In this book we present the correct ways to analyse data in appropriate situations, but it is also important that the conclusions made from the data and the analyses are valid and relevant.

When we make a determination on a substance in the laboratory we seek reassurance about a batch from a small quantity from that batch; we do not make every possible determination we could from the batch. In a consumer study we obtain evaluations from a panel of people; we do not seek opinions from everyone in the consumer population. In these and other situations where we obtain data, we base our conclusions on a **sample**.

There are many reasons for sampling. It may be to reduce the effort that would be required if we selected every item produced. It may be because the sampling is destructive (for example, tensile testing) or that the items would then be unfit for use (food). Either way, sampling is an essential feature of process-based industries.

However, when we take a sample it must be for a purpose. It should be followed by measurement, statistical analysis and a decision.

The decision must not be about the sample but the group of items, material or people from which the item was drawn. We refer to this as the **population**.

Thus the starting point for sampling must be the population about which some inference needs to be made.

Let us now look at four examples.

Statistical Methods in Practice: for Scientists and Technologists   R. Boddy, G. Smith,
© 2009 John Wiley & Sons, Ltd

What a lottery!

The key to assessing the validity of sampling is **random sampling**, although practically it can rarely be applied.

A lottery is typical of random sampling, although its purpose is certainly not sampling.

In the lottery draw every number has an equal chance of being chosen. Furthermore, the selection of a number (say, 36) will not alter or lead to any bias in the chance of any other number being chosen. When the next number is drawn, 1 or 37, for example, all numbers will have equal probabilities of being chosen.

No can do

Fizico Ltd produce many thousands of cans of drink per year. They receive the cans in pallets of 1000 cans; a typical delivery will be 20 pallets.

From each delivery they use a sampling inspection scheme to test whether the cans will leak when subjected to pressure. The test uses a higher pressure than is used in the filling process.

To select the sample they use random numbers to choose a pallet and then random numbers again to choose the position within the top layer of the pallet.

They test a sample of 100 cans and find one that leaks.

What can we say about the **population** and the **sampling**?

The sample results can be applied to the population: 'In the delivery it is estimated that 1% of cans will be defective when tested at the high pressure.' Clearly another sample may give a different result, so we do well to give a margin of error on the 1%. This is the subject of a later chapter.

The population is well defined. It is the **delivery**. Furthermore, a decision will be made about the delivery **on the basis of the sample**.

The sampling method contains two elements – choosing the pallet and choosing the cans from the top layer – both of which use random sampling. They always choose the top layer of the pallet. This should not be a problem since there should be no pattern within the pallet. Thus that layer should be representative of the pallet, so is equivalent to randomly sampling the whole pallet.

The only problem would be with an unscrupulous supplier who knows Fizico's sampling method and never puts doubtful cans on the top layer.

In this situation it is safe to assume that we have a valid sampling method.

Nobody is listening to me

Midlands Radio have a regular programme about motoring, and one edition featured safe driving using the speed limit. There is a government proposal to instal in new cars sensors which can receive input from GPS on speed limits. The sensor then sends a message to the engine management system, limiting the speed of the car to within the limit. The host of the programme has asked listeners to phone in and say whether or not they are in favour of the proposal.

Altogether 612 listeners phone to take part in the survey, of whom 511 are against the proposal.

Thus the presenter concludes that 'in the Midlands 83% are against having speed limiters in cars'.

Should we have any doubts about this conclusion?

Let us first consider the **population**. This seems undefined. Is it car drivers? Is it adults? Does it refer to the geographical area classed as the Midlands or the coverage area of the radio station? Does it just refer to radio listeners?

If we do not define the population it is not possible to ascertain whether the sample is representative.

However, even if the population was well defined, the sample is clearly not representative since it is self-selecting and as such is almost certainly biased.

Thus the sample fails on two accounts – a poorly defined population and a biased sample.

It may make good entertainment but is certainly not good science!

How clean is my river?

The biological oxygen demand (BOD) and chemical oxygen demand (COD) are measures that indicate that a river is able to support aquatic life. An agency is assessing the quality of water in an urban river and has decided to sample the water.

One of their inspectors takes three measurements. All three samples give similar values for BOD and similar values for COD. On the basis that the values showed little variability and were sufficiently low the agency concluded that the river can support aquatic life.

Let us now consider the situation. To maintain aquatic life it will be necessary that BOD and COD values are low at all times. The sampling procedure was a snapshot of the quality at one time. Thus it was representative of the quality of river water only at the sampling site at a stated time. It is well known that the quality will depend on discharges from industrial and effluent plants and these will occur spasmodically and at different times. It is therefore important that the river is sampled over many different hours, days and weeks, and at different sampling sites, to ensure that the sampling is truly representative of the river quality.

Discussion

We have emphasised the importance of considering the **population**. In later chapters we shall be using **significance tests**. All significance tests make **inferences** about the population, and it is important that the population is well defined as well as considering whether the sample is valid before any inferences are made from the data.

2

What is the true mean?

Introduction

We are often faced with data and are required to make judgements about it. Before making a judgement, take a good look at the data. Useful graphs are the **blob diagram** (or dot plot) or **histogram** for data in one variable, or the **scatter diagram** to show how two variables are related. To make an informed judgement, we also need to summarise the data. Suitable summary statistics are the **mean** (to indicate the central location of the data) and **standard deviation** (for the spread). Other useful summary statistics are the **median** and **range**. This chapter concerns itself initially with ways of presenting and summarising data before then proceeding to methods of estimating these measures of a population (for example a batch of product, or a process, or a method) from a sample of data.

Presenting data

The Deepdown Mining Company have a long-term contract to supply coal to Bordnahome Power Station. The coal is supplied by the train-load. Nine samples are taken by Bordnahome from each train, analysed for parameters such as sulfur and ash content and tested for calorific value. We shall only concern ourselves with calorific values.

The latest train-load to arrive at the power station contains 1100 tonnes of coal and has been sampled to give the following calorific values (cal/g):

46.0 54.6 58.2 50.4 42.4 50.6 48.2 52.3 49.1

Statistical Methods in Practice: for Scientists and Technologists   R. Boddy, G. Smith,
© 2009 John Wiley & Sons, Ltd

These data are variable and Bordnahome are concerned with paying a large quantity of money to Deepdown on such variable data. They would like to know what is the true mean calorific value of the train-load.

We shall concern ourselves with this question later. First let us look at the data using the simplest of diagrams to show a small set of data in one variable – a 'blob' chart in which each blob represents an **observation**. (Note: 'observation' is the general word which is used for 'measurement', 'determination', 'assessment', etc.) It is shown in Figure 2.1.

The blob chart is a simple but powerful way of expressing data. It is part of 'descriptive statistics' in many statistical packages (perhaps as a 'dot plot') but is so simple to draw. Mark out an axis, indicate the scale, and place a blob to indicate the value of each data point. The pattern of the data is easy to see.

With the coal data the diagram tells us that there are no unusual patterns. The data are distributed symmetrically, with most of the values near the centre, and there are no outliers.

Figures 2.2–2.4 show some other blob diagrams of hypothetical data which have patterns that would alert us to some sampling and analysis problems.

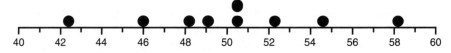

Figure 2.1   Blob diagram of the coal data

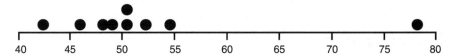

Figure 2.2   Hypothetical example containing a rogue result (outlier)

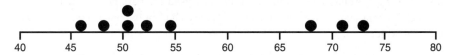

Figure 2.3   Hypothetical example containing two populations

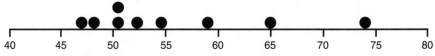

Figure 2.4   Hypothetical example with more data on the left-hand side than on the right-hand side (skewed distribution)

It does not take long to look at Figure 2.2 to decide that there is an outlier. One value is well away from the rest of the data which are clustered around a central value. Would you come to such a quick conclusion if you were just presented with a set of numbers? The cause of the outlier would need to be investigated. This is the subject of Chapter 7.

What is happening in Figure 2.3? Three outliers? It looks as though we have data from two populations. Perhaps samples were taken from two batches, or there have been assessments from consumers with different judgements. In any case, investigation is needed, but again the pattern has immediately alerted us to a problem.

In Figure 2.4 the pattern is not symmetrical. There are some kinds of data which may have this pattern, but if you are expecting a symmetrical pattern from your data this could be an indication that a process or method is out of control.

We shall see the benefits of histograms and scatter diagrams in later chapters.

Averages

There is little doubt that the coal data are straightforward compared with the hypothetical examples, and we can bear this in mind when choosing summary statistics. There are two widely used averages: the mean and the median.

The **sample mean** is obtained by summing the observations and dividing by the sample size. If we need to use a formula:

$$\bar{x} = \frac{\sum x}{n}$$

where

$\bar{x}$ is the sample mean,

$\sum x$ is the total of the observations, and

n is the sample size (or number of observations).

In the example

$$\sum x = 451.8, \ n = 9$$

so

$$\bar{x} = \frac{451.8}{9} = 50.2 \, \text{cal/g}$$

Notice the emphasis on the word 'sample'. The value of 50.2 is only a sample mean. If we took another sample of coal from the train we would be unlikely to obtain the same mean. Our real interest, however, is the true mean for the whole train-load, referred to as the 'population' mean. We shall see later that we cannot know the true mean without every single determination that is possible from all the coal in the train, but we can estimate it from our sample data.

The **sample median** is the middle observation when all the observations are placed in order of their magnitudes:

42.4   46.0   48.2   49.1   50.4   50.6   52.3   54.6   58.2

Here the sample median is 50.4 cal/g, which is an estimate of the population median.

Note that when there is an odd number of observations there is a unique middle value. With nine observations the middle one is the fifth highest. If there were an even number of observations the median would be estimated by half-way between the two in the middle. If there had been eight values

42.4   46.0   48.2   49.1   50.4   50.6   52.3   54.6

the median would be midway between the fourth and fifth, $(49.1 + 50.4)/2 = 49.75$.

Should we choose the median or the mean? Provided the distribution is reasonably symmetric (and in particular if the data follow the 'normal' distribution, which is introduced in Chapter 5) we tend to prefer the mean, giving a better estimate because it uses the magnitudes of all the data. If, however, the distribution is skewed, the median is often preferred since it is a better indication of the centre. In the ensuing chapters we shall use the mean but equivalent statistical methods are available for the median.

Measures of variability

There are three measures in common use: the range, standard deviation and relative standard deviation.

The **range** is the difference between the largest and smallest observations. For the coal data the range is

$$58.2 - 42.4 = 15.8 \, \text{cal/g}$$

The range is easy to calculate and understand but suffers from two main disadvantages:

(i) It tends to increase with sample size, which makes it difficult to compare ranges which have arisen from different sample sizes.
A new observation added to a sample will either be within the range of the sample, leaving the range unchanged, or will be outside the range of the sample, thus increasing the range. The range cannot decrease with more data, only stay the same or increase.

(ii) It depends upon only two observations and is therefore highly affected by rogue observations.

If an outlier is present, it will either be the highest or lowest value in the sample, and will contribute to the calculation of the range.

Despite these disadvantages the range is widely used in such applications as process control in which small samples of the same size are taken regularly.

The **standard deviation** is the most commonly used measure of variability. It is now easily obtained with a function on a spreadsheet or statistical package, but to show what it represents we show in Table 2.1 how it is calculated from first principles.

The first column lists the data, along with their total and the sample mean which was calculated earlier.

In the second column the sample mean has been subtracted from each observation to show the 'deviation from the sample mean'. Thus the first observation, 46.0, has a deviation of $46.0 - 50.2 = -4.2$, the minus sign indicating that it is below the mean. The second observation, 54.6, had a positive deviation, indicating it is above the mean. Notice that since these are deviations from the mean of the data, the deviations sum to zero.

In the third column the deviations are squared. They all become positive, and are added together to produce the 'sum of squares' or more correctly the 'sum of squares of deviations from the mean'.

The formula for the standard deviation includes the number of 'degrees of freedom', which in this example is one less than the sample size (number of observations).

Table 2.1 Calculation of standard deviation

	Observation	Deviation from Mean	(Deviation)^2
	46.0	−4.2	17.64
	54.6	4.4	19.36
	58.2	8.0	64.00
	50.4	0.2	0.04
	42.4	−7.8	60.84
	50.6	0.4	0.16
	48.2	−2.0	4.00
	52.3	2.1	4.41
	49.1	−1.1	1.21
Total	451.8	0.0	171.66 (sum of squares)
Mean	50.2		

We will explain degrees of freedom later.

$$\text{Sample standard deviation } (s) = \sqrt{\frac{\text{Sum of squares}}{\text{Degrees of freedom}}}$$

$$= \sqrt{\frac{171.66}{8}}$$

$$= \sqrt{21.4575}$$

$$= 4.63 \, \text{cal/g}$$

The standard deviation is an average (a strange sort of average) deviation from the mean. (It is actually the 'root mean square deviation'.) Our standard deviation is 4.63 cal/g. This means that some observations will be nearer the mean than 4.63 and some will be further away.

With a 'normal' set of data it is unlikely that many observations (about 1 in 20) will be more than 2 standard deviations ($2 \times 4.63 = 9.26 \, \text{cal/g}$) away from the mean. (This statement would not apply to the hypothetical examples in Figures 2.2–2.4.) If the next train-load gave a standard deviation of 13.89 cal/g we would know that the calorific values were three times as variable as the present load.

We also notice that the standard deviation is quoted in the same units as the original data, as are the mean, median and range. An alternative measure, the **variance**, is the square of the standard deviation. This measure, therefore, does not have the advantage of having comprehensible units. We would advise against its use except in intermediate calculations.

Relative standard deviation

The **relative standard deviation (RSD)** represents the standard deviation expressed as a percentage of the mean. It is also known as the **coefficient of variation (CV)** or **%SD**.

This is easily calculated from the standard deviation (s) and the mean ($\bar{x}$):

$$\text{RSD} = \frac{100s}{\bar{x}}$$

For the coal example,

$$\text{RSD} = \frac{100 \times 4.63}{50.2} = 9.2\%$$

The relative standard deviation is popular in many industries since it is easily understood.

It has one other advantage: if a number of sets of data have widely differing means and their standard deviations are proportional to their means, the RSD is a general measure of the variability which applies at any mean level. The converse is also true: there is no advantage in using relative standard deviation if the standard deviation is not proportional to the mean.

There are also two situations in which the relative standard deviation must not be used:

(a) With a measurement scale which does not have a true zero. For example, the Celsius scale of temperature does not have a true zero, so temperature variability must always be in absolute units, represented by the standard deviation. If we used RSD we would obtain a nonsense. For example, a negative mean temperature would give a negative RSD. There are many measurements with a true zero, including those using concentration, density, weight, and speed.

(b) When the measurement scale is a proportion between 0 and 1. For example with particle counting we can express the same result as 'proportion below 10 μm' or 'proportion above 10 μm'. If we then converted the standard deviation of the proportions into RSDs we would obtain different values depending on whether we used 'below' or 'above'.

Degrees of freedom

Let us digress from Bordnahome's data and try to explain the meaning of **degrees of freedom**. If we had only the first two observations, the calculation would proceed as shown in Table 2.2.

It is surely not a coincidence that the two values for (deviation)^2 are the same. This will always happen with two observations as the mean is mid-way between them. We say we have 1 degree of freedom since there is only one value, albeit duplicated, of (deviation)^2.

Degrees of freedom is the number of observations which can be varied independently under a constraint.

With nine observations we can vary eight of them but the ninth must be such as to make the deviations sum to zero. So with Bordnahome's data the mean is 50.2, the

Table 2.2 Calculation of standard deviation to illustrate degrees of freedom

	Observation	Deviation from Mean	(Deviation)^2
	46.0	−4.3	18.49
	54.6	+4.3	18.49
Total	100.6	0.0	36.98
Mean	50.3		

standard deviation is 4.63 and the degrees of freedom associated with the standard deviation is 8.

Degrees of freedom is always twinned with the estimate of standard deviation, no matter how it is obtained.

Confidence interval for the population mean

Let us return to Bordnahome's question: 'What is the true mean?' We have established that the sample mean is 50.2 and the standard deviation from sample to sample is 4.63. From this we can put limits on the sample mean to indicate a range within which we believe that the population mean lies. This is called a confidence interval and is calculated using the formula given below.

A **confidence interval for the population mean** (μ) is given by

$$\bar{x} \pm \frac{ts}{\sqrt{n}}$$

where

$\bar{x}$ is the sample mean,
s is the sample standard deviation, and
n is the number of observations.

Notice the notation. We use Greek letters for a population parameter such as μ for the mean, and Latin letters for sample estimates such as $\bar{x}$ for the mean and s for the standard deviation. (The population standard deviation would be denoted by σ.)

The value of t is obtained using mathematical theory and can be read from tables such as Table A.2. We shall use a 95% confidence level and we already know that there are 8 degrees of freedom for our standard deviation. This gives a value for t of 2.31.

To summarise our data: $\bar{x} = 50.2$, $t = 2.31$, $s = 4.63$, $n = 9$.

A 95% confidence interval for the population mean for the train-load of coal is

$$50.2 \pm \frac{2.31 \times 4.63}{\sqrt{9}} = 50.2 \pm 3.6$$

$$= 46.6 \text{ to } 53.8 \, \text{cal/g}$$

Thus we are 95% certain that the mean calorific value of all the coal in the train is between 46.6 and 53.8 cal/g.

The '$\pm$' part of the confidence interval, ± 3.6, is known as the margin of error.

We have not yet dealt with assumptions. Although we can easily calculate many statistics, it does not mean that they are valid. If the assumptions are grossly violated, the statistics will just be nonsense and will easily mislead us. Assumptions are the subject of later chapters. However, providing Bordnahome have taken a representative sample, the assumptions will not have been grossly violated in this situation.

What can Bordnahome and Deepdown deduce from the confidence interval? The main conclusion must be that the interval is so wide that further sampling is essential. If not, Deepdown would be advised to ask for payment using the upper limit of 53.8 cal/g and, since the price is 70p per cal/g for each tonne and the train-load contained 1100 tonnes, the payment will be £41 400. Bordnahome will of course wish to use the lower limit of 46.6 cal/g, giving a payment of £35 900. Some difference!

Sample sizes

How can the interval be reduced? It is clearly impossible to reduce the variability due to coal, since this is a state of nature. It may be possible to reduce the test variability but since this is probably small in comparison with that for the coal, little gain would ensue. The main way to reduce the confidence interval is to increase the sample size.

Before we commence a calculation, let us make one point. The size of the confidence interval is dependent upon the square root of the sample size. Thus, quadrupling the sample size will only halve the confidence interval.

The first step in calculating a sample size is to decide what width of confidence interval is required. Bordnahome and Deepdown jointly decide that a desirable width is ±1.5 cal/g. The next step is to obtain an estimate of the variability. This could be obtained from similar data or production records. In this case our best estimate is that obtained previously, namely 4.63 cal/g based on 8 degrees of freedom.

We are now in a position to estimate the required sample size:

$$n = \left(\frac{ts}{c}\right)^2$$

where c is the margin of error – the '±' part of the confidence interval.

$$n = \left(\frac{2.31 \times 4.63}{1.5}\right)^2$$
$$= 51$$

Thus a sample size of approximately 50 observations is required.

Crunch output

Throughout the book the output from the analyses using the Excel-based package *Crunch*, which is available free of charge from the internet to readers, is shown.

The analysis of the coal data in this chapter is shown below.

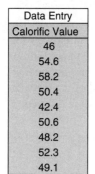
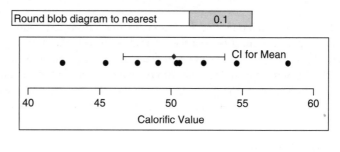

The data, with the required heading, are entered in the shaded cells.

The first feature is the blob diagram, which clearly shows how the values are distributed. The mean and confidence interval are superimposed on the graph.

The next part is situated to the right of the first part in the spreadsheet.

Summary Statistics for	Calorific Value
Mean	50.2
SD	4.63
Count	9
df	8

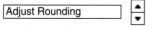
Adjust Rounding

Estimate of True Mean	
Confidence Level	95%
Margin of Error	3.56
Confidence Interval	46.64
for True Mean	to
	53.76

Sample Size Calculation	
Required Margin of Error	1.5
Required Sample Size	50.7

The summary statistics – mean, standard deviation, count (number of observations) and degrees of freedom for the standard deviation – are calculated.

The confidence interval for the mean is calculated. Notice that the confidence level can be changed to suit. The '±' part of the confidence interval, which we call the 'margin of error', is shown.

The sample size required for a margin of error of 1.5 is calculated.

How much moisture is in the raw material?

Albrew Maltsters process barley into malt for the brewing industry. To obtain a good brew it is essential that the malting barley is dried to no more than 12%

moisture. Measurement of moisture is quick and easy with an automatic probe device. The time and temperature required to reduce the moisture by x per cent is also well known. The main problem is estimating the moisture level of what is essentially a variable raw material.

Albrew take eight samples which should be representative of a batch and obtain the following moisture levels (%):

$$14.8 \quad 13.5 \quad 13.9 \quad 13.1 \quad 14.7 \quad 14.2 \quad 15.1 \quad 13.8$$

Clearly Albrew are mainly interested in the mean value since there will be a degree of mixing in the process, which should make the moisture of the final product fairly uniform. However, the mean is estimated from a sample, so we should use a confidence interval.

A 95% confidence interval for the mean is given by

$$\bar{x} \pm \frac{ts}{\sqrt{n}}$$

where
$\bar{x}$ is the sample mean,
s is the sample standard deviation, and
n is the number of observations.

We have
$$\bar{x} = 14.14, \quad s = 0.69, \quad n = 8$$

and with 7 degrees of freedom (from Table A.2 at 95% confidence level)

$$t = 2.36$$

A 95% confidence interval for the population mean moisture is:

$$14.14 \pm \frac{2.36 \times 0.69}{\sqrt{8}} = 14.14 \pm 0.58$$

$$= 13.56\% \text{ to } 14.72\%$$

Thus we are 95% certain that the moisture in the batch is between 13.56% and 14.72%.

To be on the safe side, Albrew set the drier so that it can reduce the moisture level from 14.72% to the required level of 12.0%. However, they realise that by doing so on every batch, they will, on average, be reducing batches to below 12.0% because their calculation of heat required includes the margin of error of 0.58%. If they reduce the margin of error they will reduce the cost of drying.

Could the margin of error be halved, to 0.3%?

The required sample size to give a margin of error of ±0.3% is

$$n = \left(\frac{ts}{c}\right)^2$$

where c is the margin of error – the '±' part of the confidence interval.

$$n = \left(\frac{2.36 \times 0.69}{0.3}\right)^2$$

$$= 30$$

Thus 30 samples are required.

Albrew must now decide whether the effort required to take 30 samples will more than offset the savings in heat from having a smaller margin of error.

Crunch output

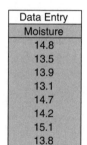

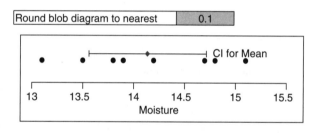

Summary Statistics for Moisture	
Mean	14.1
SD	0.69
Count	8
df	7

Adjust Rounding	▲
	▼

Estimate of True Mean	
Confidence Level	95%
Margin of Error	0.58
Confidence Interval	13.56
for True Mean	to
	14.71

Sample Size Calculation	
Required Margin of Error	0.3
Required Sample Size	29.6

Problems

2.1 Happy Cow creameries require that the fat content of incoming milk should have a mean fat content of at least 32 g/l. Seven samples from an incoming tanker-load gave the following results:

Sample	1	2	3	4	5	6	7
Fat content (g/l)	33.4	30.8	31.6	32.2	29.8	31.2	32.6

(a) Calculate the sample mean and sample standard deviation.
(b) Estimate the mean fat content of the delivery of milk by obtaining a confidence interval for the true mean.
(c) Does the delivery meet Happy Cow's requirement?
(d) How many samples would be required so that the confidence interval for the mean was no wider than ±0.5 g/l?

2.2 Tobacco is dried on a band-drier, the effect of which depends both on the speed and temperature of the drier. It is desired to operate the drier at a higher speed and temperature in order to increase throughput. An experiment is carried out at the two settings of speed and temperature using parts of the same consignment of tobacco. Samples are taken and the following results for moisture (%) are obtained:

High speed/high temp.	11.0 10.7 13.1 13.9 12.9 12.5 11.9 12.4
Normal speed/normal temp.	12.8 13.4 11.4 11.1 10.6 12.4 12.9 11.7 13.2 11.5

(a) For each speed, calculate the sample mean, sample standard deviation and a 95% confidence interval for the population mean.
(b) Examine the confidence intervals and decide whether there is any evidence that the two sets of conditions give different moisture levels.
(c) Has the experiment been successful?

2.3 Insulite blocks, which are well known in the building trade for their lightness and insulation properties, are produced using furnace ash from power stations. The ash is only a waste product from producing electricity and, as such, is not subject to any quality control. Stickash Ltd, the manufacturer of Insulite blocks, has found that the sulfur content greatly affects the manufacture and it is necessary to change the production conditions for each batch depending on its level. Furthermore, the sulfur content varies within each delivery and therefore the following sample scheme has been devised:

> Obtain 12 samples using a probe sampler. Mix the samples thoroughly and then take one sub-sample for analysis of sulfur content.

The quality of batches has occasionally been poor and Dr Small, the quality control manager, believes this could be due to the sub-sample giving a poor estimate of sulfur content. He therefore decides that each of the 12 samples from delivery D724 is to be analysed separately. (This procedure is of course too costly to institute on a regular basis.) The samples gave the following content (g/kg):

44.5 29.0 55.0 39.0 58.5 43.0 52.0 33.5 49.5 36.5 48.0 45.5

(a) Calculate a 95% confidence interval for the population mean of delivery D724.
(b) How many samples would be required to give a 95% confidence interval of ±2.0 g/kg?
(c) What has Dr Small learned from his investigation?

2.4 Crosswell Chemicals have obtained a long-term contract to supply triphenolite to a paint manufacturer who has specified that the viscosity of the triphenolite must be between 240.0 and 250.0 seconds (flow time in the viscosity cell). The first 50 batches of triphenolite produced by Crosswell give the following viscosities:

Batch Number				
1–10	11–20	21–30	31–40	41–50
240.50	245.73	248.21	242.16	247.52
245.21	247.32	243.69	251.24	244.71
246.09	244.67	246.42	241.74	246.09
242.39	244.56	242.73	247.71	241.54
249.31	245.61	249.37	245.69	243.28
247.32	256.31	245.32	244.91	247.11
243.04	246.21	244.11	245.72	242.66
245.72	242.05	250.53	246.30	245.91
244.22	246.51	247.08	243.54	246.17
251.51	243.76	240.71	244.62	248.26

(a) Summarise the above data in a blob diagram, rounding values to the nearest 0.5 seconds. Are there any points which do not appear to follow the pattern of the other observations?
(b) Deleting any observations identified in (a), re-examine the blob diagram. Do you consider that the sample has been drawn from a population with a symmetrical distribution?
(c) What is the population referred to in (b)?
(d) If a batch is above specification it has to be scrapped at a cost of £1000. If it is below specification it can be blended with other products to give in-spec material, but at a cost of £300 per batch blended. The process mean can easily be adjusted without affecting batch-to-batch variability. Use the blob diagram to estimate what value the process mean should be targeted to in order to minimise the costs of out-of-spec production.

3

Exploratory data analysis

Introduction

Before plunging headlong into a statistical analysis, it is essential to examine the data.

Methods for exploring data will allow us to gain a feel for the distribution of the data, for example:

(i) Are the values distributed in a symmetrical fashion about the centre?
(ii) Are they skewed, with a greater concentration at one end than at the other?
(iii) Are there extreme values present?

We have already seen the advantages of visually examining data using blob diagrams. These are extremely useful with small data sets but are not so useful with large data sets.

More useful with larger quantities of data is a **histogram** in which we can look at the shape of the distribution of data and make appropriate inferences.

An alternative to the histogram is the **box plot**, which is extremely useful when looking at trends in several data sets or, for example, when comparing several machines. The box plot gives a clear graphical display of the spread of the data, and in particular the median, the magnitude of the variability and any outliers.

Let us start by producing a histogram.

Histograms: is the process capable of meeting specifications?

To illustrate the use of histograms we shall use an example involving process capability.

Statistical Methods in Practice: for Scientists and Technologists   R. Boddy, G. Smith,
© 2009 John Wiley & Sons, Ltd

Trubearings Ltd make self-lubricating nylon bearings for the motor industry and need to achieve specifications laid down by Hiroll, their major customer, which is dissatisfied with the variability in the external diameter of the bearings.

Trubearings realise that the process has many parameters which can affect the process, such as different screw extruders, different feedstocks, temperature and back pressure. Any or all of these could have a considerable effect, but Trubearings' initial analysis must look at past data.

They take a random sample of 120 bearings and measure their external diameter. They find that 10 bearings are outside the specification which is 180 ± 5 mm. This is completely unsatisfactory. The process is deemed to be incapable.

Clearly Trubearings' quality is not satisfactory, but the data does not yield much information. We must examine the data more closely. The actual measurements are given in Table 3.1.

One of the best ways to examine the data is with a histogram.

The first step is to obtain the frequency, or number of measurements, in different ranges. We will group the data into five intervals.

The data are counted using 'tally marks'. A '|' is entered in the appropriate interval for each measurement and instead of a fifth '|' a diagonal line is superimposed on the previous four.

The procedure is shown in Table 3.2.

Table 3.1   External diameters of 120 bearings (mm)

179.6	183.9	181.3	183.9	182.6	176.1
181.0	180.7	182.0	179.0	182.8	178.2
181.6	179.9	181.5	179.9	183.6	181.7
180.9	184.0	183.0	177.8	181.7	182.7
179.0	187.5	180.5	180.9	187.4	180.5
183.1	179.5	182.9	183.5	179.1	181.9
182.8	179.7	183.2	181.7	180.1	178.5
183.4	180.1	181.6	181.9	182.7	182.8
181.8	187.3	179.3	181.9	182.8	179.0
186.1	184.7	180.7	182.5	182.9	179.6
181.7	181.9	181.8	182.2	178.9	181.5
182.3	184.1	182.5	182.0	178.5	187.2
183.3	180.8	182.3	183.0	179.0	180.1
181.7	184.5	182.0	182.4	179.4	180.3
183.3	181.0	181.8	184.1	181.3	181.0
188.4	178.4	181.4	180.9	180.7	178.6
179.6	181.4	180.2	189.2	182.1	181.9
180.8	177.9	180.0	180.9	180.0	182.6
182.9	179.9	189.8	181.9	181.6	181.6
177.3	180.4	177.8	187.8	181.0	186.7

Table 3.2 Counting the frequencies

Class Interval	Tally	Frequency
176.0 to 178.9	ЖЖ ЖЖ l	11
179.0 to 181.9	ЖЖ ЖЖ ЖЖ ЖЖ etc.	62
182.0 to 184.9	ЖЖ ЖЖ ЖЖ ЖЖ etc.	37
185.0 to 187.9	ЖЖ ll	7
188.0 to 190.9	lll	3

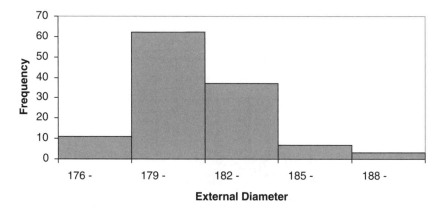

Figure 3.1 Histogram of external diameter

Table 3.2 shows the tally for each interval and the frequency. They can now be shown graphically in a histogram in which each frequency is represented by the height of each bar. This is shown in Figure 3.1.

Despite our attempts, the histogram displays little information because we have used too few groups. Clearly the size of the class interval is important. An interval of 3.0 mm is too wide. Let us repeat the procedure with a class interval of 0.2 mm. The first few frequencies are shown in Table 3.3 and the resulting histogram in Figure 3.2.

The histogram with a class interval of 0.2 mm has 70 classes. The data are very sparse, with many classes having frequencies of zero. This makes it difficult to make a judgement about the process. We need something between the class intervals of 3 and 0.2 mm.

In general, a histogram should have approximately eight classes when there are 50 data values and around fifteen when there are 100 or more data, with the class intervals being sensibly chosen as rounded numbers. As we have seen, it is not helpful to have too few or too many classes.

Table 3.3   Frequencies with a
class interval of 0.2 mm

Class Interval	Frequency
176.0 to 176.1	1
176.2 to 176.3	0
176.4 to 176.5	0
176.6 to 176.7	0
176.8 to 176.9	0
177.0 to 177.1	0
177.2 to 177.3	1
177.4 to 177.5	0
177.6 to 177.7	0
177.8 to 177.9	3
etc.	

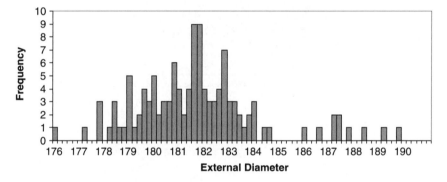

Figure 3.2   Histogram with class interval of 0.2 mm

Let us now use a class interval of 1.0 mm. The frequencies are given in Table 3.4 and the histogram in Figure 3.3.

The histogram contains 14 classes, a suitable number.

What can Trubearings deduce from this analysis?

The histogram shows that the distribution of diameters falls into two groups. The main group is within specification of 180 ± 5 mm. The specification limits are shown on the histogram. There is, however, a small group, divorced from the main group, which is outside the upper specification limit. There are two procedures – one fairly easy, one difficult – which Trubearings can undertake to achieve specification:

(i) The difficult procedure is to find the cause of the large diameters. It is important in such an exercise to ensure that items are traceable; in other words, Trubearings must know under what conditions each item was produced.

Table 3.4   Frequencies with a
class interval of 1.0 mm

Class Interval	Frequency
176.0 to 176.9	1
177.0 to 177.9	4
178.0 to 178.9	6
179.0 to 179.9	15
180.0 to 180.9	19
181.0 to 181.9	28
182.0 to 182.9	21
183.0 to 183.9	11
184.0 to 184.9	5
185.0 to 185.9	0
186.0 to 186.9	2
187.0 to 187.9	5
188.0 to 188.9	1
189.0 to 189.9	2

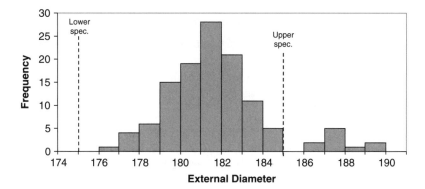

Figure 3.3   Histogram with class interval of 1.0 mm

They can then analyse the conditions under which the 10 bearings which were outside specification were produced and hopefully find the cause.

(ii) The easier procedure is to locate the main group of bearings within the middle of the specification limits – at the moment it is to the right of the middle. This would require that the mean is reduced by 0.5 mm, which could be carried out by altering an appropriate process parameter.

Box plots: how long before the lights go out?

Vi Bull is responsible for replacing lamps in the offices of Guy Fawkes Cartons. Each time a lamp fails Vi has to take a pair of stepladders, take out the old lamp

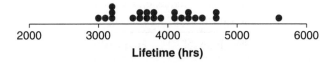

Figure 3.4   Blob diagram of lifetimes for Glowhite

and replace it with a new lamp. This all takes time, and it has been suggested that money would be saved by carrying out regular maintenance by replacing all the lamps at the same time. The period between replacements would have to be chosen carefully so that few lamps would fail before replacement, yet the interval should be sufficient to reduce costs. Clearly to calculate a replacement time we need data. Fortunately Vi is an assiduous person who has regularly recorded the date of replacing a lamp and also the time it takes her to carry out the operation. She has also recorded the manufacturer since four different manufacturers have been used. She has not recorded the wattage since all the lamps in the offices are 100 W. Let us look at the data from one of the manufacturers, Glowhite. Unfortunately only 23 lamps' lifetimes (in hours) have been recorded:

4710 3760 4050 3460 3690 3210 3240 3100 3750 3180 4720 3610
5550 4195 3930 4370 4050 3630 3680 3030 4540 4250 4260

However, 23 data points are not too many results for a blob diagram and not too few for a box plot – we would recommend a minimum of 15. Let us first look at the blob diagram in Figure 3.4.

We see from the blob diagram that the data are skewed, with many results clustered closely together around 3600 hours and then results becoming more and more widespread, with the highest value looking like an outlier. This is typical of reliability data.

The box plot is a highly graphical presentation for the distribution of a large set of data. The box plot for the Glowhite data is shown in Figure 3.5.

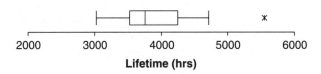

Figure 3.5   Box plot of lifetimes for Glowhite

The essential features of the box plot are as follows:

(i) The **median** – the value such that half the lifetimes are less than it, and half are greater than it.

(ii) The ends of the box represent the lower and upper **quartiles** – along with the median, they divide the population into four parts with equal numbers of observations.

(iii) The box contains the middle 50% of the population.

(iv) The length of the box is the **interquartile range**, the difference between the two quartiles.

(v) The **whiskers** continue outwards to the highest and lowest values, provided they are not 'outliers'.

(vi) **Outliers**, indicated by stars, are defined in connection with the box plot as values which are more than $1\frac{1}{2}$ times the box length beyond either end of the box.

This definition of outlier should be used only when examining a set of data empirically in a box plot; it is not the same as the formal objective definition which takes account of probabilities and the underlying distribution of the data.

To see how the summary statistics are derived, rearrange the observations in ascending order, as in Table 3.5.

Median. For a set of 23 observations, the median of the population is estimated by the middle observation, the 12th in order, 3760 hours. If there had been an even number, say 24 observations, it would have been mid-way between the 12th and 13th observations.

Lower quartile. The lower quartile is the lifetime for which there are three times as many observations above it as below it. It is estimated using the observation whose order is

$$\left(\frac{n+1}{4}\right)$$

Table 3.5   Lifetimes rearranged in ascending order

Order	1	2	3	4	5	6	7	8	9	10	11	12
Observation	3030	3100	3180	3210	3240	3460	3610	3630	3680	3690	3750	3760

	13	14	15	16	17	18	19	20	21	22	23
	3930	4050	4050	4195	4250	4260	4370	4540	4710	4720	5550

i.e. the 6th observation, which is 3460 hours. If there had been an equal number, say 24 observations, the calculation would have been $6\frac{1}{4}$ and the lifetimes would have been $\frac{1}{4}$ of the distance between the 6th and 7th observations.

Upper quartile. This is obtained using order

$$3 \times \left(\frac{n+1}{4} \right)$$

i.e. the 18th observation, which is 4260 hours.

Outliers. The box length, or interquartile range, is

$$4260 - 3460 = 800 \text{ hours}$$

The criterion for an outlier is

$$1.5 \times 800 = 1200 \text{ hours}$$

beyond the ends of the box,

less than	$3460 - 1200 = 2260$ hours
or greater than	$4260 + 1200 = 5460$ hours

There are no outliers among the low values; the only outlier is the value of 5550 hours.

Whiskers. The whisker at the lower end goes to the lowest observation, 3030 hours.

The value of 5550 hours is an outlier, so the whisker is drawn at the highest remaining observation of 4720 hours.

The box plot in practice

Vi Bull can now examine the box plot for all four manufacturers. This is given in Figure 3.6, with the number of observations.

The box plot shows that Clearview has the highest median but also the greatest variability, and that Seebrite has the lowest median and lowest variability.

However, Vi Bull is concerned about early failures since these will govern the standard replacement time. Using either the lower quartile or the low whisker, Glowhite gives the highest figure. Vi Bull decides to initially use Glowhite with 3000 hours as the replacement time but is aware that this is based on a sample of only 23 lamps and will need to make adjustments as more data become available.

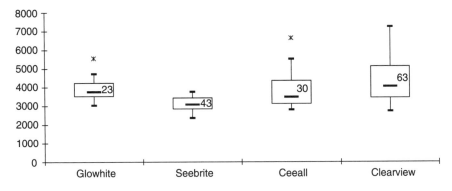

Figure 3.6   Box plot of lifetimes

Problems

3.1 The Sumoil Company is concerned about the quality of the effluent from one of its refineries. Four-hourly samples have revealed high chemical oxygen demand (COD) values. These are shown in the histogram below.

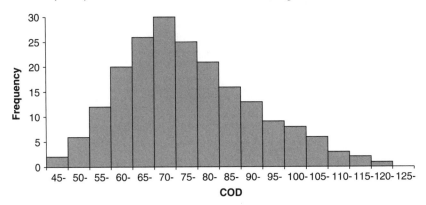

To reduce the COD values Sumoil have installed a powder-activated carbon unit. The COD values for the first 80 four-hourly samples are given below.

51.8	63.0	83.8	52.7	59.3	62.9	61.1	51.1	51.9	50.1
67.0	45.3	63.2	67.6	78.5	46.9	54.9	54.2	50.7	60.9
63.8	55.8	68.8	70.2	47.7	72.5	55.2	55.2	55.6	56.6
88.2	80.4	67.3	64.7	69.6	48.6	49.8	52.1	65.5	58.6
50.3	50.1	77.4	78.9	62.9	69.5	49.9	48.6	57.5	53.3
47.3	65.7	46.3	56.6	57.3	54.7	60.4	70.9	51.4	57.1
55.1	45.5	60.9	73.4	66.8	76.4	48.0	57.9	50.6	54.3
59.5	53.7	56.4	74.3	53.9	82.4	49.7	62.4	52.6	58.0

Produce a histogram and compare it with the histogram shown above of the results before the installation of the unit. Comment on its success or otherwise.

3.2 CDD Ltd provide computer disk drives to computer manufacturers. The management have identified a major problem with the shipping time required from receiving an order to its despatch. They have instructed Meg A. Byte to investigate the nature of the problem and suggest a course of action to reduce it. Meg first establishes that the problem is due to shortages of supplies but not due to procedural difficulties. Her next step is to analyse the data over time by computing box plots for the last 12 months. The box plots are shown below with the exception of April's figures which consisted of a sample of 19 shipping times. These are shown below (in days).

22, 12, 17, 3, 15, 15, 14, 17, 11, 20, 16, 6, 24, 22, 42, 35, 19, 7, 28.

(a) Calculate:

   (i) the median;
   (ii) the lower and upper quartiles;
   (iii) the interquartile range;
   (iv) any extreme values;
   (v) the whiskers.

(b) Add the box plot for April into the diagram below.
(c) What inference can you make from the box plots?

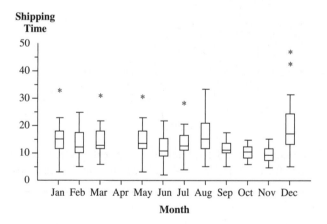

4

Significance testing

Introduction

Life is full of theories. We make theories about the weather, about people, about our sports and passions – for example, why we are slicing the golf ball when we drive from the tee. We also make theories about industrial processes and measurements. However, making decisions based on unsubstantiated theories is, to say the least, unwise. **Significance testing**, also known as **hypothesis testing**, offers us an objective method of checking theories. For example, it allows us to distinguish between changes in mean which may be due to chance and changes that are due to a cause.

In this chapter we introduce the methodology of significance testing and then proceed to test several theories involving the mean of a single sample, the use of a confidence interval to make an inference about population variability and differences in variability.

The one-sample t-test

Lodachrome Photographic Chemicals Ltd. have recently recruited a new laboratory technician, Goldwater, who having completed a training course has been let loose on analysing production samples. However, McWiley, the very cautious chief chemist, always slips a number of standard samples into the production samples to enable him to check the accuracy of the new technician. He has therefore managed to obtain the following six replicate determinations on the same sample of potassium thiocyanate which has a known concentration of 70 g/l. The values of the six

Statistical Methods in Practice: for Scientists and Technologists   R. Boddy, G. Smith,
© 2009 John Wiley & Sons, Ltd

determinations are:

$$72.1 \quad 68.0 \quad 75.3 \quad 70.6 \quad 69.9 \quad 71.9$$

$$\text{Mean} = 71.3, \quad \text{SD} = 2.463 \text{ (5 degrees of freedom)}$$

The results are plotted on the blob chart in Figure 4.1.

Figure 4.1 shows that there is nothing untoward about the data. The points are more clustered in the centre than in the tails. There is no indication of an outlier. However, the data have a high variability, with a standard deviation of 2.463. The sample mean of 71.3 is well above the true concentration, but in view of the high variability can we definitely say that Goldwater is biased? His determinations, on average, are higher than the true concentration. With this data set it is slightly difficult to decide subjectively whether he is biased and it is far better to rely on an objective method called a **significance test**, in this case the **one-sample t-test**. This procedure is as follows:

Null hypothesis   In the long run Goldwater's determinations will have a mean of 70.0 ($\mu = 70.0$).

Alternative hypothesis   In the long run Goldwater's determinations will **not** have a mean of 70.0 ($\mu \neq 70.0$).

Test value
$$= \frac{|\bar{x} - \mu|\sqrt{n}}{s}$$

where $\bar{x}$ is the sample mean,

μ is the population mean according to the null hypothesis,

$|\bar{x} - \mu|$ is the magnitude of the difference between $\bar{x}$ and μ,

s is the sample standard deviation,

and n is the sample size (number of determinations)

$$= \frac{|71.3 - 70.0|\sqrt{6}}{2.463}$$

$$= 1.29$$

Table value   From Table A.2 (t-table) with 5 degrees of freedom, 2.57 at the 5% significance level.

Decision   Since the test value is less than the table value we cannot reject the null hypothesis.

Conclusion   We cannot conclude that Goldwater is biased.

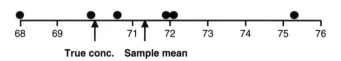

Figure 4.1   Six Determinations made by Goldwater

The conclusion is perhaps in agreement with that suggested by a visual examination of the data. In view of the large amount of variability and the small sample size it is impossible to be certain that Goldwater is biased. In other words, the difference between the sample mean of 71.3 obtained by Goldwater and the true concentration can quite easily be explained as being caused by chance. We can therefore not conclude that Goldwater is biased.

The significance testing procedure

The mechanics of a significance test, such as the one-sample t-test, are straight-forward. We obtain a test value from the data, compare it with an appropriate value from statistical tables, and reach a conclusion, for example that an analytical method is biased or is not biased. It is very easy to forget the meaning of the conclusions reached and the basis on which the test was conducted. Some comments are given below as guidance on the elements of the significance test.

(a) Experiments are conducted in order to collect data to test theories or hypotheses. The significance test allows us to test one hypothesis against another.

(b) The **null hypothesis** is a theory about a population or populations and is so called because it usually represents a 'status quo' in which there is **no change** from a specified value or no difference between populations, e.g. 'mean = 70.0'.

(c) The **alternative hypothesis** is another theory about the population or populations, representing a **change** from a specified value or a difference between populations, e.g. 'mean $\neq$ 70.0'. It does not need to be as specific as the null hypothesis.

(d) The **significance test** evaluates the data to see whether the data could reasonably have come from a population described by the null hypothesis.

(e) The **test value**, obtained from the sample data, measures the 'strength of evidence' against the null hypothesis. The larger the test value the stronger the evidence.

(f) The **table value** tells us how high the test value could reasonably be if the null hypothesis is true, i.e. 'so far and no further by chance and chance alone'.

We **reject the null hypothesis** if the sample data are unlikely, resulting in a high test value, i.e. one that would only be exceeded with a small probability. If the test value exactly **equals** the table value there would be a 5% probability of obtaining this **particular sample** (or a more extreme one) from the null hypothesis.

If the test value is **less** than the table value this particular sample (or a more extreme one) would have arisen more than 5% of the time.

If the test value is **greater** than the table value the particular sample (or a more extreme one) would have arisen less than 5% of the time.

(g) The **significance level** is the probability of wrongly rejecting the null hypothesis. The higher the test value, the smaller the probability of being wrong, and the more confident we can be about rejecting the null hypothesis and accepting the alternative hypothesis. By rejecting the null hypothesis, we are saying that chance of getting the particular sample (or one with a more extreme mean), if the null hypothesis is true, is less than 5%.

In many applications the significance level is chosen at 5%, representing an acceptable level of risk of reaching the wrong conclusion (rejecting the null hypothesis when it is true). The choice of significance level must depend on the consequences of the wrong conclusion. If it means that research will continue, then that is perhaps a reasonable risk, but if lives are endangered then a smaller risk must be appropriate.

Confidence intervals as an alternative to significance testing

In our haste to choose between hypotheses, let us not forget confidence intervals. The **confidence interval for the population mean** is given by the formula

$$\bar{x} \pm \frac{ts}{\sqrt{n}}$$

where $\bar{x}$ and n must always be obtained from the sample from which the population mean is being estimated; s is an appropriate estimate of the population standard deviation; and t is obtained from t-tables with the same degrees of freedom as the standard deviation.

The only estimate we have of the standard deviation is taken from the six determinations, has a value of 2.463 and has 5 degrees of freedom associated with it. If we had a better estimate based on more degrees of freedom we would use it.

The 95% confidence interval for the population mean for Goldwater is:

$$71.3 \pm \frac{2.57 \times 2.463}{\sqrt{6}} = 71.3 \pm 2.6$$

$$= 68.7 \text{ to } 73.9 \text{ g/litre}$$

Since this interval includes the true value of 70.0 we cannot reject, at the 5% significance level, the null hypothesis that the population mean of Goldwater's determinations could be 70.0.

If Goldwater's confidence interval had not included the true value of 70.0 we would have rejected the null hypothesis and accepted the alternative hypothesis that the population mean of Goldwater's values was not equal to 70.0.

The two statistical methods, t-test and confidence interval, always give identical conclusions, as can be shown by simple algebra; when the test value is exactly equal to the 5% table value then one limit of the 95% confidence interval will be exactly at the population mean as specified by the null hypothesis.

The confidence interval approach has the advantage that it uses real units and allows a decision to be made based on magnitude rather than a yes/no answer.

In this example, a significance test showed that there was insufficient evidence to conclude that Goldwater was biased. However, the confidence interval indicates that Goldwater is very suspect since his bias could be as much as 3.9 g/l. This figure, the greatest difference from the true value while within the confidence interval, is called the **maximum bias**. The furthest value from 70 within the confidence interval is 73.9, hence the maximum bias is 73.9 − 70.0 = 3.9.

The confidence interval approach reinforces the point that we can **never prove an equality** so we can never accept the null hypothesis.

Failure to reject the null hypothesis could be due to two reasons:

(a) Goldwater is biased but we have not obtained enough information.
(b) Goldwater really is unbiased.

Failure to obtain enough information can be rectified by taking an appropriate sample size, but before estimating the required size the chief analyst needs to know, from his practical and scientific knowledge, what size of bias he considers important. In this situation he decides that the width of the 95% confidence interval should be no more than ±1.0 g/l, which would make a bias of 1.0 g/l significant because the confidence interval would not include the true value of 70.

The sample size required to estimate the population mean within ±c using a confidence interval is given by

$$n = \left(\frac{ts}{c}\right)^2$$

Including our 'best' estimate of the standard deviation, which for Goldwater is based only on his six determinations, the calculation becomes (with $c = 1.0$; $s = 2.463$; $t = 2.57$ with 5 degrees of freedom)

$$n = \left(\frac{2.57 \times 2.463}{1.0}\right)^2$$

$$= 40$$

Undertaking 40 determinations may be impractical and McWiley must now be aware that carrying out a significance test for bias is perhaps not the best way of assessing a new technician. It is far better to compare his maximum bias with that of previous technicians.

McWiley should be aware that, if he intends to apply a significance test, 40 determinations will give him only an even chance of detecting a bias of 1.0 g/l and he would be well advised to increase the sample size above 40.

Crunch output

The analysis by the one-sample *t*-test appears in the '1-sample summary statistics' sheet of *Crunch*.

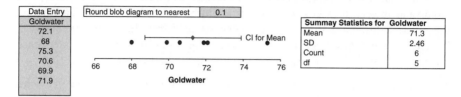

The data, with the required heading, are entered in the shaded cells. The blob diagram of the data is shown in the graph. The summary statistics – mean, standard deviation, count (number of observations) and degrees of freedom for the standard deviation – are calculated.

The next part of the display is to the right of the first part in the spreadsheet.

Estimate of True Mean	
Confidence Level	95%
Margin of Error	2.59
Confidence Interval	68.71
for True Mean	to
	73.89

1-Sample t-test	
Significance Level	5.0%
Reference Value	70
Test Value	1.29
Table Value	2.57
Decision	Not Significant
P	0.253

Sample Size Calculation	
Required Margin of Error	1
Required Sample Size	40.0

The confidence interval for the mean is calculated. Notice that the confidence level can be changed to suit. The '±' part of the confidence interval, called the 'margin of error', is shown.

The one-sample *t*-test is performed. The significance level, chosen here as 5%, can be changed to suit. The mean according to the null hypothesis, here called the 'reference value', must be entered. The test value and table values are both shown and the decision – 'significant' if the test value is greater than the table value, 'not significant' if the test value is less than the table value – is shown. The *p*-value, the probability that the null hypothesis is rejected based on that test value if the null hypothesis is true, is shown. A value less than 0.05 leads to a 'significant' decision.

The sample size required for a margin of error of 1.0 is calculated.

Confidence interval for the population standard deviation

The chief analyst is not impressed by the statement that 'we are unable to conclude that Goldwater is biased'. He thinks that Goldwater has been saved from being declared biased by his high variability. He notes that Goldwater's standard deviation is 2.463, and he decides to calculate a **confidence interval for the population standard deviation** of Goldwater's determinations.

The **confidence limits for the population standard deviation** (σ) are given by:

$$\text{Lower limit} = k_a s$$
$$\text{Upper limit} = k_b s$$

where k_a and k_b are taken from Table A.7.

For the 95% confidence interval for the population SD,

$$s = 2.463 \text{ with 5 degrees of freedom}; \quad k_a = 0.62, \quad k_b = 2.45$$

Therefore

$$\text{Lower limit} = 0.62 \times 2.463 = 1.53$$
$$\text{Upper limit} = 2.45 \times 2.463 = 6.03$$

We are 95% confident that the true standard deviation, that is, the value which we would obtain if Goldwater kept making determinations day after day, is between 1.53 and 6.03. There is clearly much uncertainty about the true precision of Goldwater.

Three points are worthy of note:

(i) The interval is very wide. We will need many determinations if the population SD is to be estimated accurately. An estimate of a standard deviation from two results, with one degree of freedom, would have such a wide interval as to make the estimate almost useless. We can see from Table A.7 that the ratio of the upper limit to the lower limit would be 70 to 1. It would need 20 degrees of freedom or more to give an interval with a ratio of 2 to 1.

(ii) The interval is not symmetrical about the sample SD of 2.463.

(iii) The interval is highly dependent on the assumption that the population follows a normal distribution. (The chief chemist considers that the population of analytical determinations is normal.) The lower limit of the 95% confidence interval for Goldwater's standard deviation is 1.53, which is higher than the value of 1.0 which he considers analysts can achieve with this method.

Crunch output

Estimate of True Standard Deviation	
Confidence Level	95% ▼
ka	0.620
kb	2.450
Confidence Interval for True Standard Deviation	1.53 to 6.03

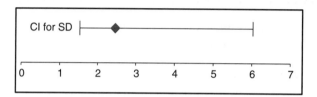

The details of the confidence interval for a standard deviation are shown, along with a graphical display of the estimate and the confidence interval. It shows dramatically how the interval is not symmetrical, unlike for the mean.

F-test for ratio of standard deviations

As further confirmation of his suspicions about Goldwater's variability McWiley decides to look at previous trials with new technicians. He selects Smith who has developed into a reasonable analyst. In his trial he carried out eight determinations, obtaining the following results:

$$69.0 \quad 70.5 \quad 70.1 \quad 69.2 \quad 71.4 \quad 70.7 \quad 70.1 \quad 70.4$$

$$\text{Mean} = 70.18, \quad \text{SD} = 0.781 \text{ (7 degrees of freedom)}$$

The blob chart in Figure 4.2 indicates that there is nothing untoward. Smith's determinations give a mean of 70.18 and a standard deviation of 0.781. We are not interested in comparing means but only in comparing standard deviations and we shall use the *F*-test to accomplish this.

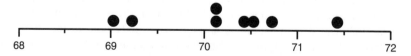

Figure 4.2   Eight Determinations made by Smith

Null hypothesis	In the long run the standard deviations of Goldwater and Smith will be equal.
Alternative hypothesis	In the long run the standard deviations of Goldwater and Smith will **not** be equal.

Test value

$$= \frac{\text{Larger SD}^2}{\text{Smaller SD}^2}$$

$$= \frac{\text{Goldwater SD}^2}{\text{Smith SD}^2}$$

$$= \frac{2.463^2}{0.781^2}$$

$$= 9.94$$

Table values Using Table A.6 (two-sided) with 5 and 7 degrees of freedom gives:
5.29 at the 5% significance level.

Decision We reject the null hypothesis.

Conclusion We can conclude that Goldwater's variability is higher than Smith's. Clearly the chief chemist is correct in his reservations about Goldwater.

Crunch output

Mean	71.3	70.2
SD	2.46	0.78
Observations	6	8
df	5	7

Significance Level	5.0%

F-test for Standard Deviations	
Test Value	9.94
Table Value	5.29
Decision	Significant
P	0.009

Data Entry	
Goldwater	Smith
72.1	69
68	70.5
75.3	70.1
70.6	69.2
69.9	71.4
71.9	70.7
	70.1
	70.4

This test is available in the worksheet entitled *2-Sample Significance Tests*. The test value is much higher than the table value at the 5% significance level, allowing us to conclude that Goldwater is less precise than Smith. The *p*-value is less than

0.01, indicating that the difference is more marked, being significant at the 1% level.

Problems

4.1 A recent modification has been introduced into a process which will considerably lower production costs. However, it is essential that the mean strength of the chemical is 5.4 mg/l. (The variability of individual batches is not important since batches are always blended together.)

(a) Set up null and alternative hypotheses and describe, in words, the population under investigation.
(b) The first ten batches with the modification gave the following results:

$$4.0 \quad 4.6 \quad 4.4 \quad 2.5 \quad 4.8 \quad 5.9 \quad 3.0 \quad 6.1 \quad 5.3 \quad 4.4$$

Use an appropriate significance test to decide whether the mean strength has altered from 5.4.

4.2 A new analytical method has been developed by a laboratory for the determination of chloride in river samples. To evaluate the accuracy of the method eight determinations were made on a standard sample which contained 50.0 mg/l.

(a) Set up the null and alternative hypotheses defining, in words, the population that is under investigation.
(b) The determinations were:

$$49.2 \quad 51.1 \quad 49.6 \quad 48.7 \quad 50.6 \quad 49.9 \quad 48.1 \quad 49.6$$

Test whether the method is biased.

(c) Calculate a 95% confidence interval for the population mean and hence obtain the maximum bias. (The maximum bias is the difference between the true concentration and the value of the 95% confidence limit which is further away from it.)
(d) What size of sample is required to give a 95% confidence interval of ±0.5 mg/l?

4.3 Nulabs Ltd. regularly run careful checks on the variability of readings produced on standard samples. In a study of the amount of calcium in drinking water the same standard was determined on six separate occasions. The six determinations, in parts per million (ppm), were:

$$9.54 \quad 9.61 \quad 9.32 \quad 9.48 \quad 9.70 \quad 9.26$$

(a) Calculate 95% confidence limits for the population standard deviation.
(b) The British Standard for calcium in water defines the 'average' standard deviation for laboratories as 0.15 ppm. Is there any conclusive evidence that the precision of Nulabs is better or worse than average? (No further calculations are necessary).

4.4 Two methods of assessing the moisture content of cement are available. On the same sample of cement the same operator makes a number of repeat determinations and obtains the following results (mg/ml):

Method A : 60   68   65   69   63   67   63   64   66   61
Method B : 65   64   62   62   64   65   66   64   64

(a) Is one method more precise (i.e. less variable) than the other?
(b) Are there any drawbacks to this method of assessment?

4.5 Mixtup Concrete buys gravel by the lorry-load from Quickholes Ltd. On average, lorries should have a net weight of at least 20 tonnes of gravel but the managing director of Mixtup, using his intuitive flair, observes the tipping of several lorry-loads and concludes that he has been 'short-gravelled'. He details his two shift supply officers, Green and Brown, to investigate further.
Green takes a random sample of two lorry-loads, which have net weights of 19.2 and 19.6 tonnes. He reports back: 'The average of my results is well below target, therefore the MD's suspicions are correct. We must demand an extra 600 kg of gravel per delivery.'
Brown takes a random sample of seven lorry-loads with net weights of 20.3, 20.5, 19.6, 19.9, 20.0, 20.3 and 20.1 tonnes. His report states: 'The average of my results is above 20 tonnes. There is nothing to worry about.'

(a) Carry out one-sample significance tests and comment on Green's and Brown's statements.
(b) The supply manager, ever anxious to please the MD, states that Brown's results must be ignored since their range is twice as large as Green's. Carry out a suitable significance test for variability and decide if he is justified.
(c) Frustrated at the lack of agreement, the MD 'analyses' all nine data points himself, and dashes off a letter to Quickholes, in which he says: 'A recent supply audit determined that 44% of your trucks are short-loaded. We expect to receive your corrective action plan and compensation proposal by return of post.'
Upon reflection he decides to consult with the company statistician (you). Calculate a 95% confidence interval for the true mean net weight per truck from all the available data and decide whether the MD should post his letter or not.

5

The normal distribution

Introduction

In a well-known demonstration of the normal distribution, known as Galton's quincunx, a large number of seeds are stored in the hopper of a tray. A plunger is removed and the seeds are released to fall through an array of pins as shown in Figure 5.1. As they land at the bottom they form the shape of a normal distribution.

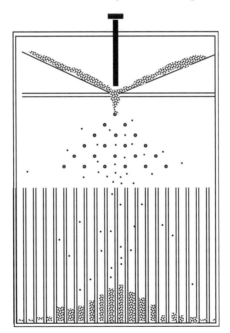

Figure 5.1   The seed tray demonstration

Statistical Methods in Practice: for Scientists and Technologists   R. Boddy, G. Smith,
© 2009 John Wiley & Sons, Ltd

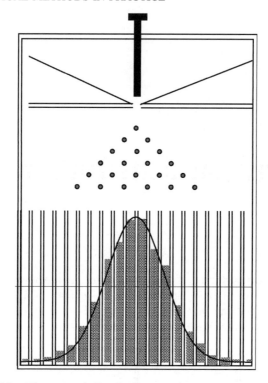

Figure 5.2   The normal distribution superimposed on the seed tray

The seed tray demonstration shows how the normal distribution occurs in practice. Figure 5.2 shows a normal distribution superimposed upon the seed tray.

The seed tray also illustrates the fundamental nature of statistics. Although we cannot predict at any time which slot the next seed will enter, we can make a statement in terms of probability, for example: 'We are 95% certain that the next seed will enter one of the 10 central slots.'

The seed tray demonstrates that if a measurement is subject to a large number of errors which are additive, the values will form a **normal distribution**.

Properties of the normal distribution

A normal distribution has the following properties:

• The mean is in the centre of the distribution.

- It has no theoretical limits but the probabilities decline rapidly in the tails.
- The population mean (μ) and population standard deviation (σ) completely define its properties. Thus we can **standardise** any calculation involving the normal distribution by subtracting the mean and dividing by the standard deviation. The standardised value is referred to as a z-**score**.

Example

Let us look at two calculations using the following example.

A survey showed that the systolic blood pressure of men between the ages of 40 and 45 has a mean of 129 mmHg and a standard deviation of 15.1 mmHg. The distribution of blood pressure within this age group is normal.

What is the probability of a male between 40 and 45 having a blood pressure of greater than 140 mmHg?

Whenever we do any calculations involving values or probabilities of the normal distribution, it is invaluable to draw the curve and indicate what is known.

In this example we know the mean and standard deviation and we are interested in the region at and above 140. We put this information into Figure 5.3.

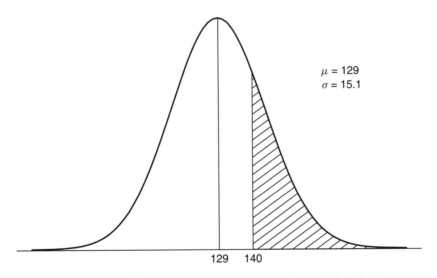

Figure 5.3   Obtaining probabilities from the normal distribution

To obtain a probability we first obtain a z-score for the value of 140:

$$z = \frac{x - \mu}{\sigma}$$

$$= \frac{140 - 129}{15.1}$$

$$= 0.73$$

Table A.1 gives the probabilities for the shaded area corresponding to any z-value. Since we are interested in probabilities on only one tail of the curve we use the table for one-tailed probabilities.

From Table A.1 (one-tailed) we see that a value of $z = 0.73$ gives a probability of 0.2327. In other words, approximately 23% of this age group of the population has a blood pressure of 140 or higher.

A z-score is simply an expression of any value in terms of multiples of the standard deviation above or below the mean.

The *Crunch* output for this example is as follows.

	Known	Calculated			Known	Calculated			Known	Calculated	
Lower x Limit			⬍	Mean	129	129	⬍	Upper x Limit	140	140	⬍
P (Obs < x)			⬍	SD	15.1	15.1	⬍	P (Obs > x)		0.233	⬍
z(lower)			⬍	Total P		0.2332	⬍	z(upper)		0.73	⬍
				Alternative Expressions of P				Capability Indices			
				P (%)		23.316%		Sigma Rating		2.23	
				P (ppm)		233161	⬍	Cpk		0.24	

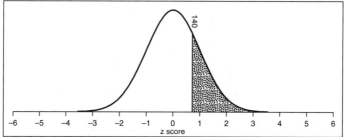

We now pose another question. Between what limits do 95% of a population lie?

This is a very practical question which is required for the setting or meeting of specifications, for example in the manufacture of a component with a specified diameter of 15 mm and specified tolerance limits. If it is permitted that no more than 5% of items may be outside the specification limits, this method will show if the specification can be met.

It is first necessary to obtain a value of z. This time there are limits on both sides of the mean, so we use the table for two-tailed probabilities. Using T1 (two-tailed)

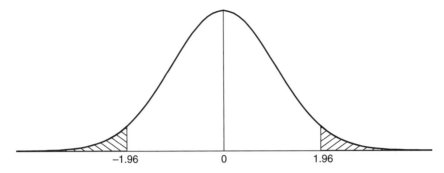

Figure 5.4   Calculating specification limits

percentage points, for 5% in the tail (referred to as significance level in the tables), $z = 1.96$ (see Figure 5.4).

To convert the limits back from z-values to values of blood pressure, we use the reverse of the calculation we used to derive z-values.

This time, starting from the z-value, we multiply it by the standard deviation and add the mean:

$$x = \mu \pm z\sigma$$

$$= 129 \pm (1.96 \times 15.1)$$

$$= 129 \pm 29.6$$

i.e. 99 to 159 mmHg

Thus 95% of the age group will have a blood pressure between 99 and 159 mmHg.

For this question the *Crunch* output is given below.

	Known	Calculated			Known	Calculated			Known	Calculated	
Lower x Limit		99.4	⬍	Mean	129	129	⬍	Upper x Limit		158.6	⬍
P (Obs < x)	0.025	0.025	⬍	SD	15.1	15.1	⬍	P (Obs > x)	0.025	0.025	⬍
z(lower)		−1.96	⬍	Total P		0.05	⬍	z(upper)		1.96	⬍

Alternative Expressions of P		Capability Indices	
P (%)	5.000%	Sigma Rating	3.46
P (ppm)	50000	Cpk	0.65

Setting the process mean

Goodstart Ltd produce packets of breakfast cereal with a nominal weight of 750 g. There is variability in the weights, which have a normal distribution with a standard deviation of 10 g. One requirement the company must meet is that no more than 2.5% of packets contain less than 735 g (Figure 5.5). The company has to take care in setting the process mean to meet this regulation.

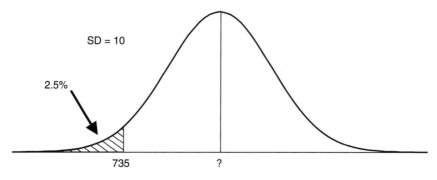

Figure 5.5   Setting the Process Mean

From Table A.1 (one-sided), if there are 2.5% of packets less than 735 g, or a probability of 0.025 of a packet having a weight below that limit, then 735 g corresponds to a z-value of -1.96.

The process mean needs to be set at 1.96 times the standard deviation above 735 g, i.e. at

$$735 + (1.96 \times 10) = 735 + 19.6 = 754.6 \text{ g.}$$

Crunch output

	Known	Calculated	
Lower x Limit	735	735	
P (Obs < x)	0.025	0.025	
z(lower)		−1.96	

	Known	Calculated	
Mean		754.6	
SD	10	10	
Total P		0.025	
Alternative Expressions of P			
P (%)		2.500%	
P (ppm)		25000	

	Known	Calculated	
Upper x Limit			
P (Obs > x)			
z(upper)			
Capability Indices			
Sigma Rating	3.46		
Cpk	0.65		

Checking for normality

Given only piece of sample data it is impossible to prove that a distribution is normal. (In fact it could be argued that the normal distribution cannot occur practically since it has no bounds – it ranges from minus infinity to plus infinity.) However, our only concern is whether the normal distribution is appropriate in a given circumstance. In particular, the problem can be troublesome with small samples.

For instance, if we are concerned with 95% tolerance intervals for individual measurements (see Chapter 6), it is important that the normal distribution is a good fit in the centre of the distribution but we need not be concerned with the extreme tails. However, if we were involved with nuclear safety and referring to probabilities of 1 in a million, we would need to know that the normal distribution was valid for tail areas of less than 0.000001 – clearly a difficult, if not impossible, task.

Let us now examine a graphical method for checking whether a sample could have been drawn from a population that could reasonably be described by the normal distribution. The data we shall use consist of the yields and levels of impurity of 15 batches of a pigment. The values are reproduced in Table 5.1.

Using the yield data, the first step is to rank the data from smallest to largest yield then extract the population percentages given in Table A.5. These values are shown in Table 5.2 and indicate, for example, that an estimated 50% of the population of batches have a yield less than 71.2, while an estimated 4.5% of batches have a yield less than 64.3. To check for normality we plot the estimated population percentage against yield on graph paper which has a vertical axis with a non-linear scale. The scale has been especially chosen to correspond to a cumulative normal

Table 5.1   Yield and impurity
for 15 batches of a pigment

Batch	Yield	Impurity
1	69.0	1.63
2	72.1	5.64
3	64.3	1.03
4	74.1	0.56
5	68.1	1.66
6	71.2	1.90
7	71.2	7.24
8	71.0	3.63
9	74.0	1.52
10	72.4	0.42
11	67.6	2.10
12	76.7	1.07
13	69.0	2.64
14	70.8	11.31
15	78.0	2.19

Table 5.2   Estimated population percentages
for the yield data

Estimated Percentage in Population	Rank	Yield
4.5	1	64.3
11.0	2	67.6
17.5	3	68.1
24.0	4	69.0
30.5	5	69.0
37.0	6	70.8
43.5	7	71.0
50.0	8	71.2
56.5	9	71.2
63.0	10	72.1
69.5	11	72.4
76.0	12	74.0
82.5	13	74.1
89.0	14	76.7
95.5	15	78.0

distribution with the result that it is squashed in the centre and wide apart at top
and bottom. We also notice that the scale contains neither 0% nor 100%, since
both values are impossible with a normal distribution.

The special scale of normal probability paper is shown in Figure 5.6.

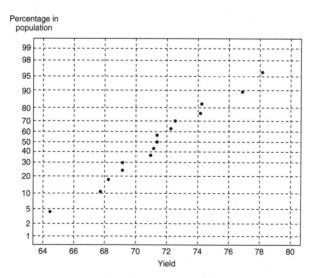

Figure 5.6   Normal probability plot of yield data

If we plotted normal distribution probability values from Table A.1 we would obtain a perfect straight line. Unfortunately the yield data contain sampling variation so all we can assess is whether a straight line is a reasonable fit. In the case of the yield data it is quite likely that the normal distribution is reasonably valid to use in the central 90% of the distribution, but with only 15 samples it would be foolish to make a statement about higher percentages.

Let us now look at the impurity data which, after ranking, is shown in Table 5.3.

Table 5.3 and Figure 5.7 refer to the impurity data and there can be no doubt that, even with a sample of size 15, impurity does not follow a normal distribution since the probability plot is highly curved. If we needed to calculate a tolerance interval – which is highly dependent upon the assumption of normality – (see Chapter 6) we could **not** use Table A.4 unless we can **transform** the data. There are many transformations – square root, arcsine, reciprocal – but we will apply one of the most useful by taking logarithms (to base 10) of all 15 impurity values. These are given in the right-hand column of Table 5.3.

Figure 5.8 shows that the logarithmic transformation has given a set of data which is a fairly close fit to a straight line and it would be reasonably valid to use the normal distribution with log-impurity, provided that we do not stray too far into the tails of the distribution.

Table 5.3   Estimated population percentages for impurity data

Estimated Percentage in Population	Rank	Impurity	Log Impurity
4.5	1	0.42	−0.38
11.0	2	0.56	−0.25
17.5	3	1.03	0.01
24.0	4	1.07	0.03
30.5	5	1.52	0.18
37.0	6	1.63	0.21
43.5	7	1.66	0.22
50.0	8	1.90	0.28
56.5	9	2.10	0.32
63.0	10	2.19	0.34
69.5	11	2.64	0.42
76.0	12	3.63	0.56
82.5	13	5.64	0.75
89.0	14	7.24	0.86
95.5	15	11.31	1.05

Percentage in
population

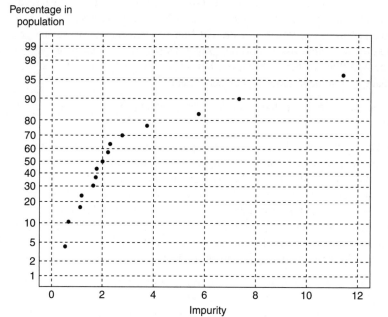

Figure 5.7   *Normal probability plot of impurity data*

Percentage in
population

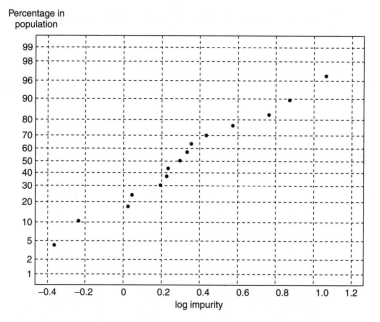

Figure 5.8   *Plot of log-impurity data*

Uses of the normal distribution

The normal distribution can be used for setting specifications. Having obtained a reliable estimate of the variability, the specification is set relative to a target mean using calculations illustrated in the above example.

The most commonly required values of the normal distribution are shown in the summary table at the foot of Table A.1. This shows that 95% of values are within ± 1.96 standard deviations of the mean (or approximately 2), 99% within ± 2.58 (approximately $2\frac{1}{2}$) and 99.8% within ± 3.09 (approximately 3). The 95% and 99.8% limits are widely used in process control.

Another use of the normal distribution is in checking that a process or analytical method is in control. Using a normal probability plot, we can check whether a set of data is non-normal. If it is non-normal, it means that a few errors are dominant and the process (or method) should be rectified to remove their effects.

We can also use the normal distribution to decide whether a significance test (or confidence interval) is valid. Some tests or confidence intervals – those involving standard deviation or individual observations – are not valid for non-normal distributions. Other tests – those involving the calculation of a mean – are valid providing a reasonable sample size has been taken.

Finally, for a highly skewed distribution, it is useful to check whether a transformation (e.g. taking logs) will convert it into a normal distribution. This is useful in extending the application of statistical methods – setting specifications and statistical tests – to non-normal distributions.

Problems

5.1 Imperial Distillers supply many duty-free outlets with their 10-year-old Glenclose malt whisky. Because of specifications imposed by their customers, and of Customs and Excise regulations, they regularly monitor the strength of their whisky.
The alcoholic strength is normally distributed with a mean of 37.5% alcohol and a standard deviation (bottle-to-bottle variation) of 0.3%.

(a) What is the **normal range** (the range enclosing the central 95% of the distribution)?
(b) A spot inspection by an excise officer gives a strength of 36.69%. How often should such a value or less occur?
(c) Does this value give cause for concern?

5.2 The British Pharmacopoeia states that the weight of active ingredient for each tablet should be within 15% of the nominal weight. Thus with a nominal weight of 200 μg the specification limits are 170 μg to 230 μg. It can be assumed that the active ingredient of the tablets forms a normal distribution.

(a) If the population standard deviation is 12 μg and the population mean weight of a batch is 200 μg, what percentage of tablets will be outside specification?

(b) If the population standard deviation remains at 12 μg but the population mean is 210 μg, what percentages of tablets will be outside specification? We shall infer that 'all tablets' in the specification really means 99.9%.

Using a population mean of 200, between what symmetrical limits will 99.9% of the tablets lie if:

(c) The population standard deviation is 12 μg?
(d) The population standard deviation is 8 μg?

What population SD is required to give 99.9% of the tablets within the Pharmacopoeia limits of 170 to 230:

(e) With a population mean of 200 μg?
(f) With a population mean of 210 μg? (Think carefully and draw a diagram.)
(g) Taking into account all the above calculations what population standard deviation should the company strive to obtain in order to meet specification?

6

Tolerance intervals

Introduction

The statement '95% of the population lies within 2 standard deviations of the mean' is a very useful rule of thumb to show the variability of a population of observations. Even with the exact coefficient of 1.96, it is true only if:

(a) the observations are normally distributed;
(b) the population mean is known exactly;
(c) the population standard deviation is known exactly.

If the mean and standard deviation are estimated from a sample of observations, a tolerance interval gives an estimate of the range encompassed by a particular percentage of the population.

As the limits of this interval are themselves estimates, we also have to state our degree of confidence in them.

Example

Eatowt Pharmaceuticals produce an anti-ulcer drug. For each batch the release limits must be within ±5% of the specified dose level. Eatowt have interpreted this requirement to mean that the **99% tolerance interval (with 95% confidence)** for tablets from a batch should be between 95% and 105% of the specified potency level.

Assays on 50 tablets from a batch of tablets of the drug gave a mean assay of 100.6 and a standard deviation of 1.5.

Statistical Methods in Practice: for Scientists and Technologists   R. Boddy, G. Smith,
© 2009 John Wiley & Sons, Ltd

A **99% tolerance interval (with 95% confidence)** for the population of single tablets produced in the batch is

$$\bar{x} \pm ks$$

where $\bar{x}$ is the sample mean, k is a coefficient from Table A.4, and s is the sample standard deviation.

In this example, we have

$$100.6 \pm (3.13 \times 1.5) = 100.6 \pm 4.7$$

$$= 95.9 \text{ to } 105.3$$

Thus we are 95% certain that 99% of tablets have a potency between 95.9 and 105.3.

The upper limit is outside the specification. The batch cannot be released.

To estimate a tolerance interval we do need a lot of data. It is very difficult to make an accurate statement about the limits for 99% of the population if we have, for example, only six assays. Thus, examining Table A.4, we notice that with only six assays the value of the coefficient would have been 5.78 instead of 3.13, resulting in an interval which not only reflects the inherent variability in the tablets but also the uncertainty due to poor estimates of the mean and standard deviation.

Confidence intervals and tolerance intervals

There is much confusion between:

(i) tolerance intervals for individual items and
(ii) confidence intervals for the mean.

In the above example a sample of 50 tablets would give a 95% confidence interval for the mean of

$$100.6 \pm 0.43$$

while the 99% tolerance interval (with 95% confidence) is

$$100.6 \pm 4.7$$

The first tells us that we know the mean fairly precisely, but the tolerance interval is far wider since it includes the inherent variation from tablet to tablet.

7

Outliers

Introduction

The idea of an 'outlier' appeals to the practical scientist/technologist. It seems intuitively obvious to him or her that a set of data may contain one or more observations which do not 'belong' to the rest of the data. We will therefore define an outlier as an observation which does not fit the pattern of the rest of the data or fit the pattern of previous sets of data.

Picking out observations which 'do not appear to belong' is a very subjective process and, when practised by the unscrupulous (or the weak), could lead to undesirable results. Fortunately statisticians have devised significance tests which can be used to determine whether or not an 'apparent outlier' really is beyond the regular pattern exhibited by the other observations. Tests for outliers, like all other significance tests, are carried out using a purely objective procedure. They can, nonetheless, be abused by persons who either:

(a) make use of outlier tests only when it suits their purpose to do so;
(b) are unaware of the assumptions underlying outlier tests;
(c) reject outlier(s) indiscriminately, without first seeking non-statistical explanations for their cause.

Underlying every test for outliers is the usual assumption that the sample was selected at random. In addition, there is an assumption concerning the distribution of the population from which the sample was taken. This second assumption is not appreciated by everyone who makes use of outlier tests, and ignorance of this assumption can lead to the drawing of ridiculous conclusions.

There are many outlier tests, each designed for a specific purpose, so this subject would indeed fill a weighty textbook. We shall however, consider only one – Grubbs' test – to illustrate the use of such tests.

Statistical Methods in Practice: for Scientists and Technologists   R. Boddy, G. Smith,
© 2009 John Wiley & Sons, Ltd

Grubbs' test

Two parameters which are important in considering the performance of the Digozo manufacturing plant are the dye-strength of the pigment and the amount of impurity in a batch. Data have been recorded from six batches in one run, as shown in Table 7.1.

We shall first consider the observations for dye-strength by looking at a blob diagram (Figure 7.1) and test whether the value of 98.2 (which is furthest from the mean) is an outlier.

Test value
$$= \frac{|x - \bar{x}|}{s}$$

where x is the most extreme observation in terms of its deviation from the mean,

$\bar{x}$ is the sample mean,

and s is the sample standard deviation.

(Notice that the mean and standard deviation are calculated including the possible outlier.)

$x = 98.2, \bar{x} = 101.1, s = 2.05$ with 5 degrees of freedom

So the test value becomes:

$$\frac{|98.2 - 101.1|}{2.05} = 1.41$$

Table value   From Table A.3:
1.89 with 5 degrees of freedom at the 5% significance level.

Decision   Since the test value is less than the table value, we cannot reject the null hypothesis.

Conclusion   We are unable to conclude that the dye-strength of the six batches contains an outlier.

Table 7.1   Dye-strength and impurity data

Variable	Determinations						Mean	SD
Dye-strength	103.5	100.4	101.3	98.2	103.3	99.9	101.1	2.05
Impurity	1.63	5.64	1.03	0.56	1.66	1.90	2.07	1.82

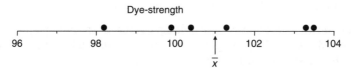

Figure 7.1   Six dye-strength observations

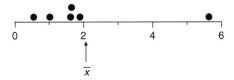

Figure 7.2   Six impurity observations

You will have noticed that Grubbs' test was carried out without a null hypothesis or an alternative hypothesis. Obviously this is unwise for a significance test of any kind. We will return to this point later.

Let us now repeat the exercise with impurity, with the observations shown in Figure 7.2, to test the value of 5.64.

Test value $= \dfrac{|5.64 - 2.07|}{1.82}$

$= 1.96$

By applying Grubbs' test to this set of data, we would conclude that the value of 5.64 was an outlier (test value $= 1.96$, table value 1.89).

Perhaps before taking any action it would be wise to correct the earlier omission, by looking more closely at the assumptions of Grubbs' test.

The null hypothesis and alternative hypothesis for Grubbs' test are:

Null hypothesis   All the observations come from the same population, **which has a normal distribution**.

Alternative hypothesis   All but one of the observations come from the same population, which has a normal distribution, whereas the most extreme value does not belong to this population.

What do we know about the distribution of dye-strength and impurity values?

Fortunately Digozo have data on many batches and the distributions of these two parameters are shown in Figure 7.3.

We can see in Figure 7.3 that dye-strength is close to a normal distribution, while clearly impurity is highly skewed. Grubbs' Test is highly dependent upon the assumption of normality and it was totally invalid to use it for impurity.

However, the shape of the distribution of impurity values suggests that a log transformation would result in a normal distribution. Grubbs' test could then be applied to the log-impurities.

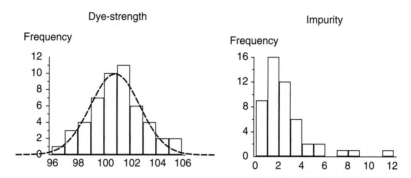

Figure 7.3   Histograms of dye-strength and impurity for 50 batches of Digozo

The log-impurity values are

$$0.21, 0.75, 0.01, -0.25, 0.22, 0.28$$

These have mean 0.20 and standard deviation 0.33, so the test value is

$$\frac{0.75 - 0.20}{0.33} = 1.65$$

By applying the Grubbs' Test in the correct situation, to the transformed data which have a normal distribution, we conclude that the most extreme observation is not an outlier.

Warning

Before using an outlier test you should ask a number of questions:

1. What pattern do the data usually follow – for example, is it a Normal distribution?
2. Are you expecting extremes only at one end of the distribution or could they occur at both ends?
3. Can you have more than one outlier?

The answers to these three questions will alter your choice of test. Grubbs' test only applies to the situation clearly specified in the hypotheses, one possible outlier in a set of data which are normally distributed. For situations with other distributions, or more than one outlier, different tests are available.

8

Significance tests for comparing two means

Introduction

In Chapter 4 we introduced **significance testing** as a method for deciding whether a set of results could be classed as beyond chance and therefore due to some prescribed event. In this chapter we introduce the **t-test** for comparing two sample means. The means can arise from many situations. In the first example described in the chapter they are from a well-designed experiment; in the second example they are from plant data.

Example: watching paint lose its gloss

The Toobright Paint Company are renowned, or so their marketing says, for the shininess of their white paint for external wood surfaces. Their research department have produced a new additive which they believe will make the paints shine even more durably. They have now decided to carry out an evaluation using the following procedure:

- Two paints are evaluated – their Hi-gloss paint (paint A) and the Better gloss paint with the additive (paint B).
- Twenty-four wooden panels are selected. These are randomly split into two groups of 12. Each panel is pre-treated, then given two undercoats before one group of 12 is painted with paint A and the other with paint B. The same painter is used throughout and the order in which he paints the panels is to a certain degree mixed so the groups are not painted sequentially.

Statistical Methods in Practice: for Scientists and Technologists   R. Boddy, G. Smith,
© 2009 John Wiley & Sons, Ltd

Table 8.1   Brightness after 48 months

Paint	Results											Mean	SD
A	86	74	78	81	90	83	70	81	85	81	87	76	81.0	5.77
B	79	85	95	83	81	88	93	83	86	89			86.2	5.12

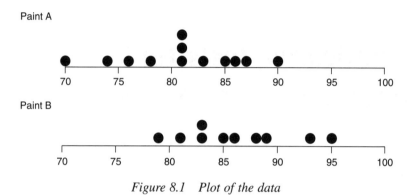

Figure 8.1   Plot of the data

- The panels are then positioned so that they are equally exposed to the elements. At selected periods the panels are cleaned and measured for brightness using a spectrometer.

The brightness values after 48 months are listed in Table 8.1 and plotted in Figure 8.1. Note that there are only 10 results for paint B because two of the panels for that group were broken during the trial when a holding pin sheared.

Clearly there are no outliers or other abnormalities in the data. Paint B appears to give higher values than paint A, although there is a considerable overlap.

We can apply the significance testing procedure in a similar manner to that shown in the previous chapter.

The two-sample *t*-test for independent samples

Null hypothesis   The population means for paints A and B are equal ($\mu_A = \mu_B$).

Alternative hypothesis   The population means for Paints A and B are **not** equal ($\mu_A \neq \mu_B$).

(The test begins by assuming there is **no** difference, and the data must establish, beyond reasonable doubt, that there is a difference.)

Test value

$$= \frac{|\bar{x}_A - \bar{x}_B|}{s\sqrt{\dfrac{1}{n_A} + \dfrac{1}{n_B}}}$$

where $\bar{x}_A, \bar{x}_B$ are the sample means for paints A and B, respectively; $|\bar{x}_A - \bar{x}_B|$ is the magnitude of the difference between $\bar{x}_A$ and $\bar{x}_B$; n_A, n_B the number of observations; and s the combined standard deviation.

The t-test for two sample means assumes that the two populations have the same standard deviation. (If this is in doubt, the F-test shown in Chapter 4 should be used to check the assumption.) We now combine the two estimates:

$$\text{Combined SD } (s) = \sqrt{\frac{(\text{df}_A \times \text{SD}_A^2) + (\text{df}_B \times \text{SD}_B^2)}{\text{df}_A + \text{df}_B}}$$

$$= \sqrt{\frac{(11 \times 5.77^2) + (9 \times 5.12^2)}{11 + 9}}$$

$$= 5.48 \text{ with } 11 + 9 = 20 \text{ degrees of freedom}$$

The test value is therefore:

$$\frac{|81.0 - 86.2|}{5.48\sqrt{\dfrac{1}{12} + \dfrac{1}{10}}} = 2.22$$

The 'test value' measures the 'strength of evidence' that there is a real difference.

When standard deviations are combined in this way their degrees of freedom are added together.

Table value   From Table A.2 with 20 degrees of freedom:
2.09 at the 5% significance level.
The 'table value' tells us how high the test value could be by chance alone.
Since the test value is greater than the table value, the observed difference was not due to chance.

To put it in formal terms:

Decision   We can reject the null hypothesis and accept the alternative hypothesis.

Conclusion   The additive improves the durability of the shininess.

The significance test starts by assuming there is no difference between the paints, then on the basis of the data allows us to reject the null hypothesis or to fail to reject it.

The significance level is the risk that we are prepared to take of concluding that there is a difference when the true situation is that there is no difference. Most researchers use a 5% risk.

An alternative approach: a confidence intervals for the difference between population means

A confidence interval for the difference between the two population means is given by

$$|\bar{x}_A - \bar{x}_B| \pm \left(ts\sqrt{\frac{1}{n_A} + \frac{1}{n_B}} \right)$$

where $\bar{x}_A, \bar{x}_B, |\bar{x}_A - \bar{x}_B|, n_A, n_B$ and s are as before; and t is a value from Table A.2 with the same degrees of freedom as the standard deviation.

In this case the most appropriate estimate of the population standard deviation is given by the combined standard deviation. This has a value of 5.48 and has 20 degrees of freedom associated with it. The 95% confidence interval is:

$$|81.0 - 86.2| \pm \left(2.09 \times 5.48\sqrt{\frac{1}{10} + \frac{1}{12}} \right) = 5.2 \pm 4.9$$

$$= 0.3 \text{ to } 10.1$$

In some ways the confidence interval is quite alarming.

Yes, we have shown, at a 5% significance level, that the additive is an improvement.

However, the size of the improvement, as judged by a 95% confidence interval, could be as high as 10.1 or as little as 0.3.

This causes great uncertainty for Toobright's management. On one hand, to add a costly additive to gain an improvement of only 0.3 makes no business sense. On the other hand, to miss an improvement of 10.1 would be to waste a wonderful marketing opportunity. The management are appalled by the uncertainty due to the width of the confidence interval and ask why a smaller one was not obtained.

Let us see if this could be achieved.

One way would be to increase the sample size.

Sample size to estimate the difference between two means

The management state that they require a confidence interval no wider than ± 2.0.

The sample size required to estimate the difference within $\pm c$ between two population means using a confidence interval is:

$$n_A = n_B = 2\left(\frac{ts}{c}\right)^2$$

$$= 2 \times \left(\frac{2.09 \times 5.48}{2.0}\right)^2$$

$$= 65 \text{ panels for each paint}$$

Thus 130 panels would be required. A trial with this number of panels would not be approved by the management since they would regard it as being too speculative and too costly.

Crunch output

Mean	81	86.2
SD	5.77	5.12
Observations	12	10
df	11	9

Significance Level	5.0%

F-test for Standard Deviations	
Test Value	1.27
Table Value	3.91
Decision	Not Significant
P	0.730

Data Entry	
A	B
86	79
74	85
78	95
81	83
90	81
83	88
70	93
81	83
85	86
81	89
87	
76	

2-Sample t-test for Independent Samples	
Combined Standard Deviation	5.48
Test Value	2.22
Table Value	2.09
Decision	Significant
P	0.039

Confidence Interval for Difference between Means	
Confidence Level	95.0%
Least Significant Difference	4.89
Confidence Interval	0.31
for True Difference	to
	10.09

Required Sample Size Calculation	
Required Difference to Estimate	2
Required Sample Size ($n_A = n_B$)	65.3
Total Sample Size ($n_A + n_B$)	130.7

The output shows the stages of the calculations of the two-sample t-test.

The mean and standard deviation for each paint are calculated.

The F-test for comparing the standard deviations, which was referred to in the chapter, is included. In this case the standard deviations are not significantly different, so it will be valid to combine them and to carry out the t-test to compare the means.

The means are significantly different. The p-value is less than 0.05, confirming that the means are significantly different at the 5% level (or less).

The confidence interval for the difference is shown, and the calculation of sample sizes for a confidence interval of ±2.0.

A production example

Cando Limited supply the drinks industry with hundreds of thousands of cans each year. These cans are automatically welded and then automatically tested for leakages. The leakages have traditionally run at about 14 per thousand but Cando are concerned that, since they changed to a new supplier of aluminium sheet, the leakage rate has increased. They have recorded leakage rates for eight weeks with the new supplier and consequently decide to look at eight weeks prior to the changeover.

The rates are shown in Table 8.2 and plotted in Figure 8.2.

Table 8.2   Leakage rates before and after the changeover of supplier

									Mean	SD
Weeks 21–28 **Old supplier**	15.2	14.8	17.5	13.1	10.5	14.1	10.3	11.0	13.31	2.57
Weeks 29–36 **New supplier**	13.1	17.3	16.8	14.4	8.9	18.9	15.9	20.9	15.78	3.70

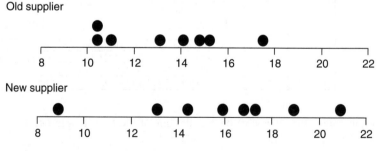

Figure 8.2   Plots of leakages before and after changeover of supplier

Again the data appear to contain no abnormalities, so we can proceed to carry out significance tests.

In the previous example we said that if the standard deviations appeared sufficiently different they should be checked by an F-test. We shall now do that.

Null hypothesis   The population standard deviations of weekly leakage rates for the old and new suppliers are equal.

Alternative hypothesis   The population standard deviations of weekly leakage rates for the old and new suppliers are **not** equal.

Test value
$$= \frac{\text{Larger SD}^2}{\text{Smaller SD}^2}$$
$$= \frac{\text{New supplier SD}^2}{\text{Old supplier SD}^2}$$
$$= \frac{3.70^2}{2.57^2}$$
$$= 2.07$$

Table values   Using Table A.6 (two-sided) with 7 and 7 degrees of freedom gives:
4.99 at the 5% significance level.

Decision   We cannot reject the null hypothesis.

Conclusion   There is no evidence that the weekly data for the new supplier are more variable than those for the old supplier.

We can therefore combine the standard deviations.

$$\text{Combined SD } (s) = \sqrt{\frac{(\text{df}_A \times \text{SD}_A^2) + (\text{df}_B \times \text{SD}_B^2)}{\text{df}_A + \text{df}_B}}$$
$$= \sqrt{\frac{(7 \times 2.57^2) + (7 \times 3.70^2)}{7 + 7}}$$
$$= 3.19 \text{ with } 7 + 7 = 14 \text{ degrees of freedom}$$

If the standard deviations had been significantly different we may not have proceeded with testing the means since the variability itself was giving important information about the process. If we still needed to test the means we would have used a modified form of the t-test for unequal standard deviations.

Let us now carry out a two-sample t-test.

Null hypothesis   The population mean weekly leakage rates for the old and new suppliers are equal. ($\mu_A = \mu_B$)

Alternative hypothesis   The population mean weekly leakage rates for the old and new suppliers are **not** equal. ($\mu_A \neq \mu_B$)

Test value

$$= \frac{|\bar{x}_A - \bar{x}_B|}{s\sqrt{\dfrac{1}{n_A} + \dfrac{1}{n_B}}}$$

$$= \frac{|13.31 - 15.78|}{3.19\sqrt{\dfrac{1}{8} + \dfrac{1}{8}}}$$

$$= 1.54$$

Table value   From Table A.2 with 14 degrees of freedom: 2.14 at the 5% significance level.

Decision   We cannot reject the null hypothesis.

Conclusion   There is insufficient evidence to say that the failure rate with cans from the new supplier is worse.

Confidence intervals for the difference between the two suppliers

We calculate the confidence interval using the formula given earlier:

$$|\bar{x}_A - \bar{x}_B| \pm ts\sqrt{\frac{1}{n_A} + \frac{1}{n_B}} = |13.31 - 15.78| \pm 2.14 \times 3.19\sqrt{\frac{1}{8} + \frac{1}{8}}$$

$$= 2.47 \pm 3.42$$

$$= -0.95 \text{ to } 5.88$$

In other words, the new supplier might be better by 0.95 per 1000 or worse by 5.88 per 1000 cans. This again shows that it is impossible to conclude which supplier is better.

Sample size to estimate the difference between two means

The difference between the means is substantial in the opinion of Cando management but is not statistically significant. In their opinion a difference of 2.0 per 1000 cans is important.

Let us estimate the sample size required to detect such a difference, which is also achieved if the confidence interval is no wider than ± 2.0:

$$n_A = n_B = 2 \left(\frac{ts}{c} \right)^2$$

$$= 2 \times \left(\frac{2.14 \times 3.19}{2.0} \right)^2$$

$$= 23.4$$

In other words, 24 weeks using the new supplier (compared with 24 weeks with the old supplier) would be needed to detect the difference of 2.0 in their performance.

Crunch output

Mean	13.3	15.8
SD	2.57	3.7
Observations	8	8
df	7	7

Significance Level	5.0%

F-test for Standard Deviations	
Test Value	2.07
Table Value	4.99
Decision	Not Significant
P	0.359

Data Entry	
Old	New
15.2	13.1
14.8	17.3
17.5	16.8
13.1	14.4
10.5	8.9
14.1	18.9
10.3	15.9
11.0	20.9

2-Sample t-test for Independent Samples	
Combined Standard Deviation	3.19
Test Value	1.54
Table Value	2.14
Decision	Not Significant
P	0.144

Confidence Interval for Difference between Means	
Confidence Level	95.0%
Least Significant Difference	3.42
Confidence Interval	-0.96
for True Difference	to
	5.88

Required Sample Size Calculation	
Required Difference to Estimate	2
Required Sample Size ($n_A = n_B$)	23.4
Total Sample Size ($n_A + n_B$)	46.8

Conclusions

The example highlights a number of difficulties.

First of all, Cando would need 24 weeks to evaluate a new supplier. This is far too long in practical terms and we must ask ourselves the reason for this. It is due to high week-to-week variability. With high variability and a small sample it will always be difficult to establish that a difference is significant.

This brings us to a second point. The management would be concerned if the failure rate increased by 2.0 with the new supplier. However, the week-to-week failure rate ranges from 8.9 to 20.9 per 1000. They would be able to make a far bigger impact on performance if they could find the reasons for such a variable performance rather than concerning themselves with the small differences between suppliers.

Thirdly, we must ask about the validity of the comparison. It is always possible to compare two sets of numbers and decide whether they are significantly different or not. The validity of doing so requires judgement on the part of the researcher. In this example there must be considerable doubt about its validity since the trial was carried out over 16 weeks in which it is possible that many process or personnel changes may have taken place and it is also extremely likely that environmental conditions changed. The only way of making this valid would be to intersperse production runs using sheets from the old and new suppliers.

Problems

8.1 Two batches of an alloy were tested to determine whether their melting points (in °C) were significantly different. Altogether eight samples were taken from the first batch and 12 from the second. The results were as follows:

Batch	Results						n	Mean	SD
A	571.6	574.7	574.2	569.6	570.8	572.4	8	572.55	1.776
	573.7	573.4							
B	574.1	570.6	573.3	569.6	572.3	571.4	12	571.40	1.827
	568.9	573.6	568.7	571.2	572.7	570.4			

(a) Calculate the combined standard deviation.
(b) Carry out a significance test to determine whether the batches have different means.
(c) Obtain a 95% confidence interval for the difference between the two means.
(d) How many samples are required to obtain a 95% confidence interval for the difference of ±1.2 °C?

8.2 Jackson and Jackson deliver products to a large number of retailers. As part of their quality improvement programme they carry out customer satisfaction surveys. These are comprised of many questions referring to reliability, finish, returns, display units, delivery and telephone queries. Each answer is placed on a scale of 1 to 5 and they are then combined to a final score which lies between 0 and 100. Each quarter they select 15 retailers at random. The scores for the last two quarters are shown below. (Notice that some retailers failed to reply.)

Quarter			Scores					n	Mean	SD
1st	85	81	88	92	76	78	93	81	13	84.23	5.39
	79	84	86	82	90						
2nd	69	82	92	76	78	83	87	67	11	78.09	8.09
	73	69	83								

(a) Is there any evidence that the variability in scores is different in the two quarters? (Use an appropriate significance test.)

(b) Is there any evidence that the mean scores are different? (Use an appropriate significance test.)

(c) Calculate a 95% confidence interval for the difference between the mean scores for the two quarters.

(d) Jackson and Jackson would like to ascertain any significant trends in performance quickly and therefore would like a margin of error for the difference between means of ±3.0. How many retailers would they need in order to achieve this?

9

Significance tests for comparing paired measurements

Introduction

We have already seen how significance tests can be used to make objective decisions and procedures for carrying out one-sample and two-sample t-tests.

In this chapter we look at the paired-sample t-test, which is a valuable addition to our selection of significance tests. We shall see that its use is limited, but when it can be applied, it is very efficient.

Comparing two fabrics

Synlene is a new synthetic fibre with wonderful properties, apart from its dyeing characteristics which make deep shades difficult to obtain. Previously researchers have developed a pre-treatment using Alphamate which is undoubtedly successful in creating deep shades but is very expensive. However, a new compound, Betamate, has emerged from a competitor, and is being offered on a basis which would result in substantial cost savings.

It was decided, therefore, to carry out an experiment to compare the two pre-treatments on eight fabrics. Each fabric was split in two, with one half being pre-treated with Alphamate, the other half with Betamate. The results are given in Table 9.1.

Statistical Methods in Practice: for Scientists and Technologists   R. Boddy, G. Smith,
© 2009 John Wiley & Sons, Ltd

Table 9.1   Darkness of fabrics

									Mean	SD
Alphamate	78	92	97	82	90	93	104	76	89.0	9.67
Betamate	73	85	95	83	81	88	98	69	84.0	9.93

The data could be analysed in two ways:

(i) correctly, by using the paired-sample t-test, or
(ii) incorrectly, by using the two-sample t-test for independent samples which we met in Chapter 8.

To illustrate the importance of choosing the correct test, let us first analyse the data in the wrong way, using the t-test for independent samples.

The wrong way

Null hypothesis   The population means for Alphamate and Betamate are equal ($\mu_A = \mu_B$).

Alternative hypothesis   The population means are **not** equal ($\mu_A \neq \mu_B$).

Test value

$$= \frac{|\bar{x}_A - \bar{x}_B|}{s\sqrt{\dfrac{1}{n_A} + \dfrac{1}{n_B}}}$$

where $\bar{x}_A, \bar{x}_B$ are the sample means for Alphamate and Betamate, respectively; $|\bar{x}_A - \bar{x}_B|$ is the magnitude of the difference between $\bar{x}_A$ and $\bar{x}_B$; n_A, n_B the number of observations; and s the combined standard deviation.

The t-test assumes that the two populations have the same standard deviation. We combine the two estimates:

$$\text{Combined SD } (s) = \sqrt{\frac{\left(\text{df}_A \times \text{SD}_A^2\right) + \left(\text{df}_B \times \text{SD}_B^2\right)}{\text{df}_A + \text{df}_B}}$$

$$= \sqrt{\frac{\left(7 \times 9.67^2\right) + \left(7 \times 9.93^2\right)}{7 + 7}}$$

$$= 9.80 \text{ with } 7 + 7$$
$$= 14 \text{ degrees of freedom}$$

The test value is therefore:

$$\frac{|89.0 - 84.0|}{9.80\sqrt{\dfrac{1}{8} + \dfrac{1}{8}}} = 1.02$$

Table value   From Table A.2 with 14 degrees of freedom: 2.14 at the 5% significance level.

Decision   We cannot reject the null hypothesis.

Conclusion   We have insufficient evidence to conclude that the treatments give different shades.

The test value is 1.02, compared with a table value at a 5% significance level with 14 degrees of freedom of 2.14. So despite a difference of 5.0 between the sample means, the conclusion is that there is not a significant difference between Alphamate and Betamate.

This is of concern. Synlene know that a difference of 5.0 would be costly in terms of dyeing, yet we have concluded that there is no significant difference.

Synlene would like to detect a difference of 3.0, which is equivalent to a confidence interval of ± 3.0. How many samples would they require? Let us use the calculation from Chapter 8. The sample size required to estimate the difference within $\pm c$ between two population means using a confidence interval is:

$$n_A = n_B = 2 \left(\frac{ts}{c} \right)^2$$

$$= 2 \times \left(\frac{2.14 \times 9.80}{3.0} \right)^2$$

$$= 98 \text{ fabrics for each treatment}$$

Crunch output

Mean	89	84
SD	9.67	9.93
Observations	8	8
df	7	7

Significance Level	5.0%

F-test for Standard Deviations	
Test Value	1.06
Table Value	4.99
Decision	Not Significant
P	0.945

Data Entry	
Alphamate	Betamate
78	73
92	85
97	95
82	83
90	81
93	88
104	98
76	69

2-Sample t-test for Independent Samples	
Combined Standard Deviation	9.8
Test Value	1.02
Table Value	2.14
Decision	Not Significant
P	0.325

Confidence Interval for Difference between Means	
Confidence Level	95.0%
Least Significant Difference	10.51
Confidence Interval	−5.51
for True Difference	to
	15.51

Required Sample Size Calculation	
Required Difference to Estimate	3
Required Sample Size ($n_A = n_B$)	98.2
Total Sample Size ($n_A + n_B$)	196.4

The paired sample t-test

Clearly an experiment involving 98 fabrics for each treatment is out of the question and we must ask ourselves if there is a way in which we could achieve the same conclusion in fewer experiments. In order to do this we first turn our attention to the standard deviation. This comprises two sources of variability:

(a) The trial-to-trial variability. Even if we could use the same fabric twice we would not obtain the same results.

(b) The fabric-to-fabric variability. This was taken into account in the design of the experiment by splitting each fabric in two, but was not taken into account in the analysis. We should have used a paired-sample t-test to analyse the data.

We can now carry out a paired-sample t-test using the stages of a significance test as given in Chapter 4.

The experiment is 'paired' because for each observation on Alphamate from part of a fabric, there is an observation on Betamate from the other part of the same fabric. Using this knowledge, the results can be displayed as in Table 9.2.

For each fabric we have subtracted the Betamate result from the Alphamate result to form a difference. We therefore have now a sample of eight differences from a population of differences and we can use the **paired t-test** to see if the mean difference is not equal to zero. In examining Table 8.2 we notice that the standard deviation of the differences is greatly reduced from that of the original results, thus vindicating the decision to use a matched-pair design. We also notice that seven of the eight differences are positive (Alphamate gave higher readings).

We now proceed to use the paired t-test on the differences.

Null hypothesis   The population means for Alphamate and Betamate are equal, i.e. the mean of the population of differences is equal to zero ($\mu_d = 0$).

Alternative hypothesis   The mean of the population of differences is **not** equal to zero ($\mu_d \neq 0$).

Test value $= \dfrac{|\bar{x}_d - \mu_d|\sqrt{n_d}}{s_d}$

where
$\bar{x}_d$ is the sample mean of differences,
μ_d is the mean difference according to the null hypothesis (0),
s_d is the sample standard deviation of the differences,
n_d is the number of differences.

$= \dfrac{|5.0 - 0|\sqrt{8}}{3.16}$

$= 4.47$

Table 9.2   Results from a paired-sample experiment

Fabric	A	B	C	D	E	F	G	H	Mean	SD
Alphamate	78	92	97	82	90	93	104	76	89.0	9.67
Betamate	73	85	95	83	81	88	98	69	84.0	9.93
Difference	5	7	2	−1	9	5	6	7	5.0	3.16

Table values From Table A.2 (t-table) with 7 degrees of freedom:
2.36 at the 5% significance level;
3.50 at the 1% significance level.

Decision- We can reject the null hypothesis at the 1% significance level.

The experimenter again chose a significance level of 5%, but as the test value of 4.48 was not only greater than the table value at 5% (2.36) but also greater than the 1% table value (3.50) he can be even more sure that there is a difference between the treatments, as his risk of making the wrong conclusion is less than 1%.

Thus there is little doubt that Betamate is inferior and a decision must be made by weighing up the decrease in cost against the decrease in darkness. In doing so it is helpful to look at a 95% confidence interval for the mean difference:-

$$\bar{x}_d \pm \frac{ts_d}{\sqrt{n_d}}$$

To summarise our data:

$$\bar{x}_d = 5.0, \quad n_d = 8, \quad s_d = 3.16 \text{ (7 degrees of freedom)}, \quad t = 2.31 \text{ (5% level)}$$

A 95% confidence interval for the population mean difference is:

$$5.0 \pm \frac{2.36 \times 3.16}{\sqrt{8}} = 5.0 \pm 2.6$$

$$= 2.4 \text{ to } 7.6$$

Lastly, we can calculate the number of observations required to give a 95% confidence interval for the difference in means to within ± 3.0:

$$n_d = \left(\frac{ts_d}{c}\right)$$

$$= \left(\frac{2.36 \times 3.16}{3.0}\right)^2$$

$$= 6.2$$

Thus 7 pairs of observations are needed, from only 7 fabrics, a considerable improvement on using independent samples which required 98 fabrics for each treatment.

This shows how much more efficient paired experiments are, where it is possible to use them.

Crunch output

Alphamate	Betamate	Differences
78	73	5
92	85	7
97	95	2
82	83	-1
90	81	9
93	88	5
104	98	6
76	69	7

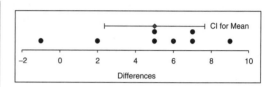

Summary Statistics for Differences

Mean	5
SD	3.16
Count	8
df	7

Adjust Rounding	▲ ▼

Paired t-test

Significance Level	5.0%
Reference Value	0
Test Value	4.47
Table Value	2.36
Decision	Significant
P	0.003

Estimate of True Mean Difference

Confidence Level	95%
Margin of Error	2.64
Confidence Interval for True Mean Difference	2.36 to 7.64

Sample Size Calculation

Required Margin of Error	3
Required Sample Size	6.2

Presenting the results of significance tests

There are many different ways of presenting the results from a significance test. Three methods of presentation are given in Table 9.3 but no recommendation is made about which one is preferable since the choice of method is mainly determined by the one currently in use in a particular industry.

Table 9.3   Methods of presenting the results of significance tests

Results	Decision	Significance Level	Probability	Stars
Test value is less than 5% table value	Cannot reject null hypothesis at the 5% significance level	Not significant	$p > 0.05$	n.s.
Test value is greater than the 5% table value but less than the 1% table value	Reject the null hypothesis at the 5% significance level	Significant at the 5% level	$p < 0.05$	* usually placed next to the test value
Test value is greater than the 1% table value but less than the 0.1% table value	Reject the null hypothesis at the 1% significance level	Significant at the 1% level	$p < 0.01$	**
Test value is greater than the 0.1% table value	Reject the null hypothesis at the 0.1% significance level	Significant at the 0.1% level	$p < 0.001$	***

One-sided significance tests

All the significance tests so far carried out have all had one thing in common. The alternative hypothesis has included the words 'not equal to'. In other words, the mean could either have increased or decreased and it could therefore have changed in either one of **two** directions. These tests are therefore known as **two-sided tests** and are perhaps best in keeping with the uncertainties of life in a practical environment in which, no matter how certain we are that our alteration to the process will result in an improvement, the opposite can often occur.

There is, however, a purist approach to statistics which is based on a scientific method. This involves:

(a) Formulating your hypothesis before you start the investigation and defining the statistical methods which are to be applied. We should **not** explore the data for abnormalities or go in search of hypotheses to explain the pattern exhibited by the data.

(b) Using scientific theory, our alternative hypothesis must state the direction of change. This is therefore referred to as a **one-sided test**.
If the sample mean is in the opposite direction to the hypothesised change it will be of no interest to the researcher and **no** significance test will be carried out.

(c) The researcher must state the significance level that is to be applied. Only one table value, often referred to as a critical value, will be obtained and a yes/no decision will be taken based on comparing the test value with this table value. We should note that for a one-sided test we use a different column in the tables than that used for a two-sided test.

Problems

9.1 Tertiary Oils have developed a new additive, PX 235, for their engine oil which they believe will decrease petrol consumption. To confirm this belief they carry out an experiment in which petrol consumption in miles per gallon is recorded for six different makes of car driven at the same speed around the same circuit. Each car is run with and without the additive. The results were as follows:

Make of Car	A	B	C	D	E	F
Oil with additive	26.2	35.1	43.2	36.2	29.4	37.5
Oil without additive	24.6	34.0	41.5	35.9	29.1	35.3

(a) Is the change in petrol consumption attributable to the additive? Has the additive significantly increased the number of miles per gallon?

(b) How many cars are needed to obtain a 95% confidence interval for the change to within ± 1.0 mpg?

(c) If the data had been analysed as a two-sample t-test for independent means (in other words, the analyst had not appreciated the paired nature of the design) how many cars would have been needed to answer part (b)?

9.2 The Research and Development laboratories of DePong Global have reached an exciting stage in the development of a new environmentally safe roll-on underarm deodorant. They are now ready to put four formulations to the test in trials involving a panel of 24 volunteers.

The method of testing is as follows. A deodorant is applied under one arm. After a fixed period of time in a hot environment an instrument, the Autosniffer, measures the strength of odour in the armpit.

DePong would like some way of making direct comparisons among all four formulations, but realise that each panellist has only two arms and hence two armpits.

Twelve of the panellists are taken at random and treated with formulation A under one arm and formulation B under the other. The other twelve are treated with C and D.

The readings from Autosniffer are given below:

Panellist:	1	Left	31	Right	34
	Formulations:	A		B	

Panellist:	2	Left	39	Right	37
	Formulations:	D		C	

Panellist:	3	Left	58	Right	58
	Formulations:	B		A	

Panellist:	4	Left	56	Right	51
	Formulations:	A		B	

Panellist:	5	Left	39	Right	53
	Formulations:	C		D	

Panellist:	6	Left	47	Right	58
	Formulations:	B		A	

Panellist:	7	Left	57	Right	59
	Formulations:	A		B	

Panellist:	8	Left	39	Right	30
	Formulations:	D		C	

Panellist:	9	Left	48	Right	59
	Formulations:	C		D	

Panellist:	10	Left	30	Right	35
	Formulations:	B		A	

Panellist:	11	Left	58	Right	44
	Formulations:	D		C	

Panellist:	12	Left	61	Right	79
	Formulations:	C		D	

Panellist:	13	Left	60	Right	46
	Formulations:	A		B	

Panellist:	14	Left	62	Right	53
	Formulations:	D		C	

Panellist:	15	Left	30	Right	36
	Formulations:	C		D	

Panellist:	16	Left	37	Right	49
	Formulations:	B		A	

Panellist:	17	Left	60	Right	50
	Formulations:	D		C	

Panellist:	18	Left	47	Right	48
	Formulations:	A		B	

Panellist:	19	Left	33	Right	52
	Formulations:	C		D	

Panellist:	20	Left	29	Right	36
	Formulations:	B		A	

Panellist:	21	Left	59	Right	60
	Formulations:	A		B	

Panellist:	22	Left	66	Right	53
	Formulations:	D		C	

Panellist:	23	Left	32	Right	43
	Formulations:	C		D	

Panellist:	24	Left	47	Right	52
	Formulations:	B		A	

(a) Carry out a significance test to determine whether there is a difference in the effectiveness of deodorants with formulations A and B.

(b) How many panellists would be required so that a mean difference of 5.0 would be significant?

(c) Carry out a significance test to compare B and D.

(d) How many panellists would be required so that a difference of 5.0 would be significant?

(e) Which of these tests was more appropriate for panel data?

(f) How might you plan the experiment so that direct comparisons could be made among all four formulations?

10

Regression and correlation

Introduction

In previous chapters we have applied statistical methods to analyse characteristics such as impurity and yield, which are often referred to as variables. All of our analyses had one common feature: they treated each variable separately and did not try to find relationships between the variables. In this chapter we examine a technique known as **regression** analysis which can be used to describe the relationship between two (or more) variables. We also introduce **correlation** analysis, which quantifies the strength of the relationship.

Obtaining the best straight line

For some time the Joint Works Committee of Reed Industries have been pressing for a bonus scheme to be introduced in the packaging department. The packaging manager has been firmly against this suggestion for the reason that orders are individually packed to a customer's requirements and as such it would be impossible to offer a standard time for the job. However, he has reluctantly agreed that Pickles, a work-study officer, should carry out a feasibility study.

As an initial step in the investigation Pickles examines the time taken to pack a consignment of filters. Although each box is of the same size the number of filters packaged varies to meet a customer's requirements. He therefore decides to study one particular operative, and record both the time taken to complete each package, in seconds, and the number of filters. Altogether the operator's performance is observed on 10 successive consignments and Pickles records the data given in Table 10.1.

Statistical Methods in Practice: for Scientists and Technologists   R. Boddy, G. Smith,
© 2009 John Wiley & Sons, Ltd

Table 10.1   Data recorded by the work-study officer

Consignment	1	2	3	4	5	6	7	8	9	10	Mean	SD
No. of filters	16	20	12	14	15	18	21	10	19	15	16.0	3.53
Packaging time	910	985	735	805	820	870	960	645	890	760	838	105.8

Pickles also observes visually that the packaging time is comprised of two elements:

(a) a fixed time element which involves obtaining a box and fastening it after inserting the filters;

(b) a variable time element which is directly proportional to the number of filters in the package.

Examining Table 10.1, we notice that consignments 5 and 10 both comprise 15 filters but the packaging times are 820 and 760 seconds, respectively. Thus packaging time is not only dependent upon the 'number of filters' but also has a random variation associated with it. However, the standard deviation of 105.8 seconds consists of both the random element and the deliberately induced variation of 'number of filters'.

A useful first step in the analysis is to plot a scatter diagram, but before doing so we must first draw a distinction between the different nature of the two variables, 'number of filters' and 'packaging time'. We refer to the packaging time as the **response** (or **dependent variable**) and the number of filters as the **independent variable**. The distinction is clear in this situation because packaging time may be dependent upon the number of filters, but by no conceivable stretch of the imagination could we believe that the number of filters in a consignment is dependent upon packaging time. It is usual to refer to the response as y and the independent variable as x.

The choice of a response and independent variable is important and, as we shall see later in the chapter, is not always straightforward.

Having chosen time as the response (y), we can plot a scatter diagram as shown in Figure 10.1. Examination of the scatter diagram reveals that there is not a perfect relationship between packaging time and number of filters, but the data appear **scattered** about a straight line, often referred to as a linear relationship. Thus although it would appear that packaging time will increase with the number of filters it is impossible to predict precisely what length of time will be required to pack a given number of filters. Thus from our analysis we require:

(a) a measure of the relationship between packaging time and number of filters, which will be shown as the best straight line;

(b) a measure of how closely the points fit the line;

(c) a measure of uncertainty of predicting future packaging times.

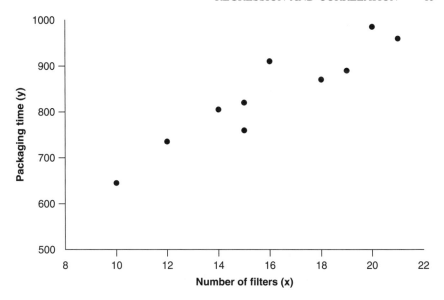

Figure 10.1   Scatter diagram of time against number of filters

We shall initially concentrate on the fitting of a straight line. The general form of the equation of a straight line is

$$y = a + bx,$$

where y is the predicted value of the response,

   a is the value of the intercept,

   b is the value of the slope,

and x is a stated value of the independent variable.

The equation of a straight line is represented diagrammatically in Figure 10.2.

If we enter the data into a statistical package, we will obtain some statistics:

   $a = 388$ (the intercept),

   $b = 28.125$ (the slope),

   $r = 0.938$ (the correlation coefficient, to be explained later).

The regression equation of the line of 'best fit' is therefore

$$y = 388 + 28.125x$$

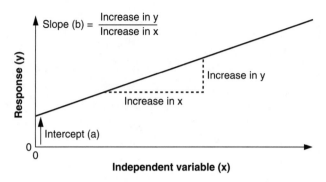

Figure 10.2   Equation of a straight line

where x is the number of filters in a consignment, and y is the predicted time in seconds to pack the consignment.

Although we do not need to know the details, the formulae are as follows:

$$b = \frac{\sum (x - \bar{x})(y - \bar{y})}{\sum (x - \bar{x})^2}$$

$$a = \bar{y} - b\bar{x}$$

$$r = \frac{\sum (x - \bar{x})(y - \bar{y})}{\sqrt{\sum (x - \bar{x})^2 \sum (y - \bar{y})^2}}$$

In Figure 10.3 the line is drawn on the scatter diagram. What do the two parameters a and b represent?

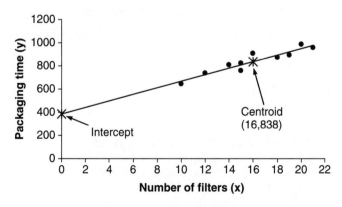

Figure 10.3   The regression line

a, the intercept, represents the fixed time element for any consignment, irrespective of the number of filters (388 seconds).

b, the slope, represents the additional time required for each filter in a consignment (28.125 seconds per filter).

The line of best fit will always go through two points on the graph:

- the intercept (the point where the line intercepts the y-axis) and
- the centroid (the 'centre of gravity' of the points, whose coordinates are the mean of x and the mean of y).

The intercept was 388 seconds.

The means of x and y were included in Table 10.1 – they were 16.0 and 838, respectively.

We notice that whereas the majority of points are near the line, a few are a considerable distance away. Returning to Figure 10.3, one feature is obvious: the data points are well away from the origin (with x-values ranging from 10 to 21 filters, whilst the origin is at 0) and we must wonder how much faith can be put on the value of the intercept and consequently how much uncertainty there is in our estimate of the fixed element in the packaging time.

By what criterion can the regression line be called the best straight line? This question can be answered by first looking at the difference between the observed packaging time and predicted packaging time, referred to as a **residual**.

The residuals are shown in Figure 10.4 as vertical lines from the points to the regression line. The residuals can also be obtained in tabular form, as given in Table 10.2.

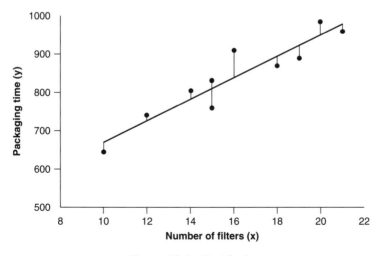

Figure 10.4   Residuals

Table 10.2   Calculation of residuals

Consignment no.	No. of filters (x)	Actual packaging time (y)	Predicted packaging time $388 + 28.125x$	Residual
1	16	910	838	72
2	20	985	951	34
3	12	735	726	9
4	14	805	782	23
5	15	820	810	10
6	18	870	894	−24
7	21	960	979	−19
8	10	645	669	−24
9	19	890	922	−32
10	15	760	810	−50

We notice that some of the points are above the line, and therefore have positive residuals, and some are below the line with negative residuals. You may have noticed something else about the pattern of the residuals – we shall return to this later.

The residuals from the fitted line can be used in exactly the same way as deviations from a mean to determine a standard deviation. In regression analysis this is referred to as the residual standard deviation (RSD). The 'best' straight line minimises the residual standard deviation and as a consequence sends the line through the centroid. Any other line would have a greater RSD.

The residual standard deviation can be obtained directly from the residuals by squaring each residual and then computing the sum of the squared residuals which is referred to as the residual sum of squares:

$$RSD = \sqrt{\frac{\text{Residual sum of squares}}{\text{Degrees of freedom}}}$$

The residual sum of squares is

$$72^2 + 34^2 + \ldots + (-50)^2 = 12\ 167$$

The degrees of freedom for the RSD are **two** less than the sample size; we 'lose' two degrees of freedom by estimating the values of a and b from the data. The sample size is 10, so there are 8 degrees of freedom. Therefore

$$RSD = \sqrt{\frac{12167}{8}} = 39.0 \text{ seconds}$$

We can interpret this in two ways:

(a) The 'average deviation' from the line is 39.0 seconds.
(b) If data were collected for many consignments, each containing the **same** number of filters, the times would have a standard deviation of 39.0 seconds. The value of 39.0 seconds should be compared with the standard deviation of the times for the ten consignments (see Table 10.1) which was 105.8 seconds. Thus fitting the regression line, thereby eliminating the effect of the number of filters on the packaging time, has had a marked effect of reducing the original standard deviation of y, and hence the uncertainty of predicting a packaging time.

Confidence intervals for the regression statistics

We can now turn our attention to the accuracy of prediction. First let us consider the regression equation. The coefficients a and b have been calculated from the sample data in Table 10.1. The ten consignments referred to in this table are only a sample from all the consignments that may have been included. Another sample would yield different coefficients. What we are really interested in is the true equation, but this is unobtainable and we must confine our attention to confidence intervals for the true equation and its coefficients.

The **confidence interval for the true intercept** is given by

$$a \pm \left(t \times \text{RSD} \sqrt{\frac{1}{n} + \frac{\bar{x}^2}{(n-1)(\text{SD of } x)^2}} \right)$$

For our data, from Table A.2, with 8 degrees of freedom, for a 95% level of confidence,

$$t = 2.31, \quad a = 388, \quad \text{RSD} = 39.0, \quad n = 10, \quad \bar{x} = 16, \quad \text{SD of } x = 3.53$$

So the 95% confidence interval is:

$$388 \pm \left(2.31 \times 39.0 \sqrt{\frac{1}{10} + \frac{16.0^2}{9 \times 3.53^2}} \right) = 388 \pm 139 = 249 \text{ to } 527 \text{ seconds}$$

Thus we say that there is a 95% chance the true value of the fixed element in the packaging time is between 249 and 527 seconds. This interval is extraordinarily wide, but this is not too surprising in view of the fact that the measurements were all made well away from the y-axis.

Let us now turn our attention to the slope.

The **confidence interval for true slope** is given by

$$b \pm \left(t \times RSD \sqrt{\frac{1}{(n-1)(SD \text{ of } x)^2}} \right)$$

So our 95% confidence interval is

$$28.125 \pm \left(2.31 \times 39.0 \sqrt{\frac{1}{9 \times 3.53^2}} \right) = 28.125 \pm 8.50$$

$$= 19.6 \text{ to } 36.6$$

Thus there is a 95% chance that the packaging time increases by between 20 and 37 seconds for each filter added to the consignment. Again this interval is very wide, reflecting both the amount of variability about the line and the small sample size. There are a number of ways of reducing the width of the confidence interval for the slope:

(i) We could obtain a larger sample size. But the benefit is only in relation to $\sqrt{n}$, so for example it will need approximately four times the sample size to halve the width.

(ii) We could reduce the residual standard deviation, if investigation of the cause of the large residual standard deviation enabled changes to be made which would make this happen.

(iii) We could increase the standard deviation of x, determining the regression line over a larger spread of numbers of filters.
We can calculate the **confidence interval for the true mean packaging time for a series of consignments containing the same number of filters** (X). This is given by

$$(a + bX) \pm \left(t \times RSD \sqrt{\frac{1}{n} + \frac{(X - \bar{x})^2}{(n-1)(SD \text{ of } x)^2}} \right)$$

Let us consider a 95% confidence interval for consignments containing 20 filters. This is given by

$$(388 + 28.125 \times 20) \pm \left(2.31 \times 39.0 \sqrt{\frac{1}{10} + \frac{(20 - 16.0)^2}{9 \times 3.53^2}} \right) = 950 \pm 45$$

$$= 905 \text{ to } 995 \text{ seconds}$$

Thus we are 95% certain that for all consignments with 20 filters the true relationship will give a packaging time of between 905 and 995 seconds. If we require to

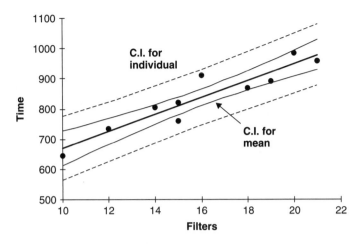

Figure 10.5   Confidence bands for the mean time and individual times

make the confidence interval apply to a range of values of the independent variable we calculate a series of confidence intervals and draw the confidence bands as shown in Figure 10.5.

We should also note that the above formula will produce a confidence interval for the intercept if we set X to zero.

Finally, we can calculate a **confidence interval for the true packaging time of a single consignment with X filters**:

$$(a + bX) \pm \left(t \times \text{RSD} \sqrt{1 + \frac{1}{n} + \frac{(X - \bar{x})^2}{(n - 1)(\text{SD of } x)^2}} \right)$$

For example, for $X = 20$ filters, we obtain

$$(388 + 28.125 \times 20) \pm \left(2.31 \times 39.0 \sqrt{1 + \frac{1}{10} + \frac{(20 - 16.0)^2}{9 \times 3.53^2}} \right) = 950 \pm 98$$

$$= 852 \text{ to } 1048 \text{ seconds}$$

We notice that the above formula is the same as the one in the previous section except an extra '1' is included under the square root. This represents the variability due to individual packages. This term cannot be reduced by carrying out a larger experiment.

The confidence bands for predicting times for individual consignments are also shown in Figure 10.5.

The uncertainty in prediction for a single consignment is somewhat wider than for the mean of all packages of the same size.

Notice that the width is narrowest near the centre of the data, as might be expected from the nature of the formula for the confidence intervals. It is more marked for prediction of the mean, where the curvature of the bands is evident, but also occurs with prediction for single consignments, although the formula is dominated by the '1' rather than the $(X - \bar{x})^2$.

Extrapolation of the regression line

The regression line fitted to the graph of packaging time versus number of filters gives a useful method of estimating the packaging time for different numbers of filters. However, if we use values of the independent variable outside the range of our experience we are implying that this linear relationship can be extended into this different range of values. In general this is a very dangerous assumption to make. If the true relationship is not exactly linear, then the slightest curvature could give results very different outside the known range.

For example, a study on a popular make of car considered the effect on petrol consumption of travelling at different speeds under motorway conditions. The consumption in the study ranged from 42 miles per gallon (mpg) at 50 miles per hour (mph) to 31 mpg at 70 mph. A regression equation,

$$y = 70 - 0.55x$$

was fitted to the data with consumption as the response (y) and speed as the independent variable (x). Now this linear equation might be a reasonable approximation to the true relationship for speeds between 50 and 70 mph. If, however, we use the equation to predict the consumption at 10 mph we obtain $70 - (0.55 \times 10) = 64.5$ mpg, whereas the true figure would be extremely low. Clearly it is dangerous to use a relationship outside the range of measured values unless we know that there is a physical relationship which is linear. We should also bear this in mind when quoting a confidence interval for the intercept since this involves extrapolation to $x = 0$ which may be well outside the range of the data.

Correlation coefficient

The regression equation gives no indication of the strength of the linear relationship between two variables or even whether there is a significant relationship. This can be achieved by obtaining the sample correlation coefficient (r), which has a value of 0.938 in our example. Before we interpret this figure let us look at some examples in Figure 10.6.

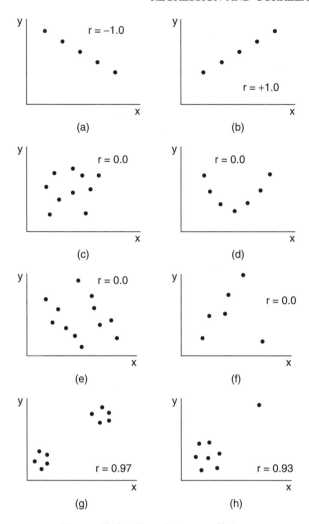

Figure 10.6   Correlation coefficients

A correlation coefficient will have a value between -1 and $+1$ inclusive. A value of -1 corresponds to a perfect negative relationship, as shown in Figure 10.6(a), whilst a value of $+1$ corresponds to a perfect positive relationship, as shown in Figure 10.6(b).

A correlation coefficient of 0 is an indication of the absence of a **linear** association between two variables (see Figure 10.6(c)), but it would be unwise to conclude that a zero correlation always implies that there is no association; Figures 10.6(d)–(f) clearly show situations in which there are strong associations but the correlation coefficients are zero. In Figure 10.6(d) there is clearly a relationship, but it is not linear. The best straight line goes neither up nor down, so $r = 0$. In Figure 10.6(e) it appears that two sets of data, each with a strong relationship, have been put on the same graph, while in Figure 10.6(f) any evidence of a relationship has been distorted by an outlier. It is equally dangerous to conclude that high correlations indicate a strong association since, as shown in Figures 10.6(g)–(h), data which fall into two groups can give high correlation coefficients, as in both examples a straight line will go close to all the points.

A further danger in the interpretation of a correlation coefficient lies in the assumption that a correlation coefficient is an indication of the existence of a cause–effect relationship between the two variables. If we were to list the annual sales in the UK of Scotch whisky alongside mean annual salaries of Methodist ministers for the past twenty years, we would find a high positive correlation between the two variables, but no one would suggest that there is a direct cause–effect relationship between the two. A voluntary reduction in ministers' salaries will not result in a reduction in the sales of whisky any more than a drop in the sales of whisky will cause a decrease in the ministers' salaries.

One point should be emphasised strongly. The correlation coefficient is a measure of the linear association between two variables and is the same **no matter which of the variables is chosen as the response**. In fact correlation is most useful when the variables cannot be categorised as either response or independent variables and also when a sample has been drawn from a population. Thus if we take a representative sample from the adult male population of the UK and find the correlation between height and chest size is 0.80, we would expect to find a similar value in the population as a whole. However, where the levels of the independent variable can be deliberately varied by the researcher, the correlation coefficient is as much dependent upon the choice of levels as on the strength of relationship. Therefore comparing correlation coefficients between different experiments is to be avoided.

Is there a significant relationship between the variables?

In many situations the scientist would be uncertain whether there is a relationship or whether the value of the sample correlation coefficient is due to chance, for even if there is no association between the two variables the calculated correlation coefficient is unlikely to be zero because of sampling variation. A simple test for the association of two variables can be carried out using the correlation coefficient. For convenience we shall use the packaging example to demonstrate this, although with this particular example there would seem little point in checking for a significant association.

Null hypothesis	Within the population of all consignments there is no linear relationship between packaging time and number of filters.
Alternative hypothesis	Within the population there is a linear relationship between packaging time and number of filters.
Test value	The magnitude of the sample correlation coefficient: $|r| = 0.938$.
Table value	From Table A.12 with 8 degrees of freedom (2 less than the sample size):
	0.632 at the 5% significant level;
	0.765 at the 1% significance level.
Decision	Reject the null hypothesis at the 1% significance level.
Conclusion	There is a relationship between packaging time and number of filters.

How good a fit is the line to the data?

The correlation coefficient is the simplest measure for assessing how good a fit is provided by the equation. The **percentage fit** is also often quoted:

$$\text{Percentage fit} = 100r^2$$

The percentage fit gives a measure of how much of the original variability of y has been reduced by fitting the line:

$$\text{Percentage fit} = 100 \times 0.938^2$$

$$= 87.9\%$$

We have therefore explained 87.9% of the original variability by fitting the number of filters. Thus 87.9% represents the reduction in the original sum of squares.

Assumptions

With the software available the computational difficulties in regression analysis will fade into insignificance. What we now need to consider are the assumptions behind regression analysis, a knowledge of which will help us to use regression analysis with safety. These assumptions are as follows:

(a) The true relationship between the response and independent variable is linear.

(b) The true value of the independent variable can be measured precisely, whilst the measured value of the response is subject to sampling error. Regression analysis is not particularly sensitive to this assumption especially with high correlations. Methods of dealing with errors in both variables have been developed.

(c) The residuals follow a normal distribution.

(d) The residuals are independent of each other.

(e) The magnitude of the residuals is of the same order for all observations and does not increase with increasing magnitude of the independent variable.

It is unlikely that all, or indeed any, of these assumptions will be met precisely, but it is essential that no one assumption is grossly violated.

Consider assumption (a). It is unlikely, for example, that the true relationship between packaging time and number of filters will be precisely linear, but a linear relationship often provides a good fit to the data even if the true relationship is curved. As mentioned previously, however, this relationship might not hold if we extrapolate outside the range of the data.

Assumption (b) is undoubtedly true since the number of filters is measured exactly but the packaging time varies from consignment to consignment even if they contain the same number of filters.

The validity of assumption (c) can be assessed with either blob diagram, histograms or a normal probability plot. We should not forget that an outlier will also render this assumption invalid and perhaps this kind of error is the most likely cause for violating the assumption. However, with this set of data further analysis reveals nothing untoward.

The validity of assumption (d) can be checked by plotting the residuals against the independent variable or, for that matter, any other variable which the researchers feel may affect the value of the response. The residuals should always form a random pattern, which certainly is not the case if we plot the residuals for packaging time against the consignment number as shown in Figure 10.7.

The residuals are not independent of each other. They have a definite pattern, decreasing steadily from a high positive residual to a high negative residual throughout the course of the trial, as may have been spotted in Table 10.2. Thus, from a statistical point of view, assumption (d) is invalid. From a work study point of view the performance of the operator is improving as the study proceeds. The reason for this must be pure conjecture, but unfortunately it must be ascertained and further work carried out. The fact that the samples are not independent means that the regression analysis is invalid and should not be used as a basis for setting up a bonus scheme.

We have now established that the packaging time is dependent upon two variables – number of filters and consignment number. The analysis of two (or more) variables can be carried out using multiple regression analysis which

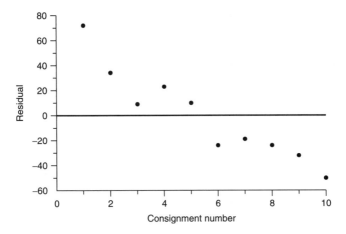

Figure 10.7   Residuals against consignment number

is a very powerful technique for examining many independent variables and determining which ones are significant, either by themselves or in combination (called an interaction) with another variable. This technique is especially useful in process improvement and optimisation studies.

Assumption (e) can be best illustrated if we consider the use of regression in calibration. To calibrate an instrument, whether used in chemical or physical analyses, it is necessary to prepare and measure several samples which have a known value. For example, the samples may have been produced by dissolving, in distilled water, weighed amounts of potassium salt. Thus the sample contains a known amount of potassium and could then be presented to an instrument which records a reading in, say, microamps. The results of such an experiment are given in Figure 10.8.

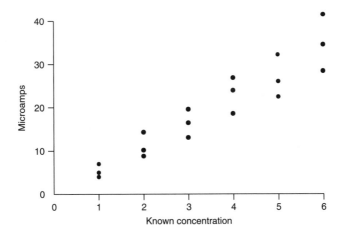

Figure 10.8   Results of a calibration experiment

Clearly assumption (e) is invalid in this setting since the magnitude of errors is definitely increasing with the magnitude of the independent variables. This is typical of calibration data and is one of a series of problems which means that calibration experiments require special statistical methods.

Crunch output

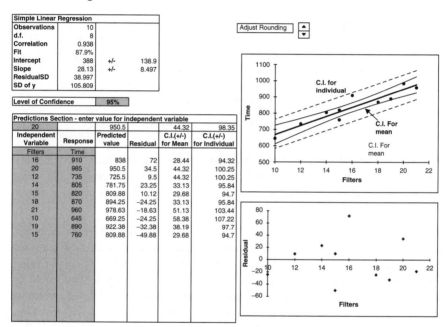

Simple Linear Regression			
Observations	10		
d.f.	8		
Correlation	0.938		
Fit	87.9%		
Intercept	388	+/-	138.9
Slope	28.13	+/-	8.497
ResidualSD	38.997		
SD of y	105.809		

Level of Confidence	95%

Predictions Section - enter value for independent variable					
20		950.5		44.32	98.35
Independent Variable	Response	Predicted value	Residual	C.I.(+/-) for Mean	C.I.(+/-) for Individual
Filters	Time				
16	910	838	72	28.44	94.32
20	985	950.5	34.5	44.32	100.25
12	735	725.5	9.5	44.32	100.25
14	805	781.75	23.25	33.13	95.84
15	820	809.88	10.12	29.68	94.7
18	870	894.25	−24.25	33.13	95.84
21	960	978.63	−18.63	51.13	103.44
10	645	669.25	−24.25	58.38	107.22
19	890	922.38	−32.38	38.19	97.7
15	760	809.88	−49.88	29.68	94.7

Problems

10.1 Culture City Electronics (Research and Development Division) have developed an instrument called Instant Count which will detect the presence and indicate the density of certain bacteria in foods. At the prototype stage, a reading is displayed which is a value of a measurable property of a solution of a sample of the food in a medium after incubation in specified conditions for a short time. It is eventually required to include a calibration equation in the software of Instant Count so that it will display the bacterial count.

To this end, Billy Dalgleish, a microbiologist in the R & D Division, prepares samples of unpasteurised milk and determines counts of listeria by Instant Count and by traditional plate counting.

The data, as Instant Count readings and logarithms of standard plate counts (SPC), are tabulated below.

Sample	Instant (x)	log SPC (y)
1	28	4.58
2	14	3.35
3	24	4.59
4	32	4.96
5	20	4.69
6	13	4.31
7	24	4.28
8	24	4.30
9	32	5.06
10	40	6.30
11	18	3.92
12	38	6.18
13	26	5.06
14	29	4.36
15	38	5.13
16	30	4.74
Mean	26.9	4.74
SD	8.16	0.744

(a) Which variable has been chosen as the independent, and which is the response? Why do you think they were chosen that way?

(b) Enter the data into the 'Regression' sheet of **Crunch** and examine the output. Do the graphs of the data and the residuals suggest anything unusual?

(c) What are the slope and intercept of the best-fit regression line?

(d) What is the value of the residual standard deviation?

(e) Carry out a significance test to determine whether there is evidence of an association between the variables.

(f) If the Instant Count reading for a sample is 25, what would the estimated log SPC be? Compute a 95% confidence interval.

10.2 Dee Seize, the Medical Officer of Health for the borough of Ayre, has long believed that the high proportion of homes with lead plumbing in the borough, coupled with the plumbosolvent nature of the water, is a contributory factor to the poor health in the borough. The local water company are, however, less convinced. Dee therefore decides to carry out a scientific study to confirm her theory and, as a first step, she collects information from 12 houses with lead plumbing on the lead levels in the tap water and lead levels in blood samples from the occupants. (In each case the blood sample was taken from the lady of the house, together with her age and length of occupancy.) The following results were obtained:

House	Blood lead (µg/l)	Water lead (µg/l)	Years of occupancy	Age of occupant
A	30	170	4	25
B	21	100	28	33
C	19	140	1	65
D	37	190	35	75
E	29	90	40	78
F	30	260	6	49
G	18	60	10	55
H	38	390	8	39
I	12	70	2	23
J	50	260	45	81
K	45	330	7	37
L	43	400	12	40

(a) Starting with blood lead and water lead:

(i) Decide which variable is the response and which is the independent variable.

Using *Crunch*, enter blood lead and water lead data into the appropriate columns of the 'Regression' sheet.

(ii) Examine the scatter diagram and the residual plot. Are there any unusual patterns?

(iii) Using the correlation coefficient, decide whether there is a significant relationship between the variables.

(iv) What would be the mean blood lead level for houses where the water lead level is 160 µg/l?

(b) Repeat (ii) and (iii) with years of occupancy replacing water lead.

(c) Repeat (ii) and (iii) with age of occupant replacing years of occupancy.

(d) Return to the relationship between blood lead and water lead.

(i) By how much would we expect that blood lead would increase when water lead increases by 1 µg/l?

(ii) Examine the width of the confidence intervals. Why has house L the largest width?

(iii) We have established that there is a relationship between blood lead and water lead but there are still some high residuals. Can we explain the size of residuals using other variables? (Correlate the residuals with years of occupancy and age of occupant.)

11

The binomial distribution

Introduction

So far we have been concerned with 'continuous' data. Such data can be quoted
to as many significant figures as are obtainable by the measuring equipment. Thus
the height of a person is usually measured to the nearest inch or centimetre but
there is no reason why it could not be recorded more accurately. In this chapter we
will look at **discrete** data. This type of data can only take certain values – usually
integers – in the form of the number of responses in two categories, such as 'fail'
or 'pass'.

Example

Nocharge plc produce electrical insulators for outdoor use which, because of their
exposure to the elements, require a high-quality surface finish. This is achieved in
the manufacture by applying a coating to the insulator and then baking the finish
in a continuous belt furnace. Every insulator is visually inspected as it leaves the
furnace for a defect of the surface called 'hazing'. The management of Nocharge
have instructed the process engineer that the number of hazed insulators should
not exceed 10%.

The process is somewhat lacking in controls, with the result that the process engineer often needs to make adjustments to it. However, he never likes to make
adjustments if the process is working satisfactorily and he has therefore devised
a procedure in which he takes three insulators as soon as he arrives for work
every morning and examines them for hazing. Should there be one or more hazed
insulators he decides to stop the process and adjust it.

Statistical Methods in Practice: for Scientists and Technologists   R. Boddy, G. Smith,
© 2009 John Wiley & Sons, Ltd

Let us suppose that the furnace is operating with conditions which produce exactly 10% 'hazed' insulators in the long run. For convenience we shall refer to hazed insulators as 'defectives'. Thus we need to know the probability of finding exactly 0, 1, 2 or 3 defectives in a sample of size 3. This problem can be represented by a probability tree as shown in Figure 11.1.

To draw the tree we consider one insulator at a time and its two possible states, 'defective' or 'effective', together with their probabilities. Since there are 10% defectives in the population, the probability of an insulator, chosen at random, being defective is 0.1. This is repeated for all three insulators. The final outcome is found by following the tree from left to right and the probability for each outcome is obtained by multiplying together the probabilities for each branch.

We can now examine the final outcomes to obtain the probabilities of 0, 1, 2 or 3 defectives in the sample. For example, the probability of obtaining exactly

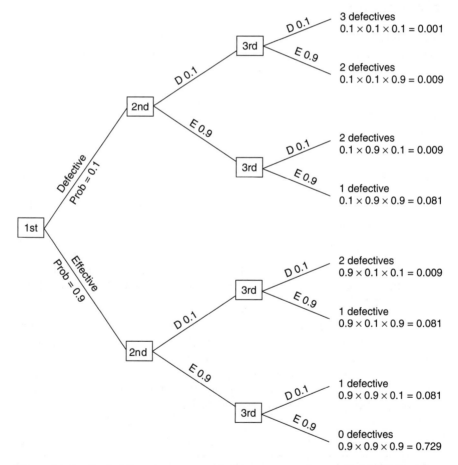

Figure 11.1   Probability tree representing inspection schemes for hazed insulators

2 defectives in the sample is $(0.009 + 0.009 + 0.009) = 0.027$. These probabilities can be summarised in the form of a probability distribution as shown in Table 11.1.

Examination of Table 11.1 shows that if the furnace is producing exactly 10% defectives there is a 0.729 (72.9%) chance that he will not adjust the process using this 'quality control scheme'. The general term used for the distribution of probabilities in Table 11.1 is the **binomial distribution**; we will have more to say about it later.

Let us now look at the binomial distribution when the furnace is operating well, with 5% defectives in the long run, and when the furnace is operating badly, with 20% defectives. The probabilities can again be calculated using a probability tree to give the values in Table 11.2, which for completeness also includes 10% defectives.

The probabilities from Table 11.2 are shown in Figure 11.2, which illustrates the fact that the binomial distribution takes many different shapes even with the same sample size. It also highlights the deficiencies in the process engineer's control scheme, since even with the furnace operating well (i.e. only 5% defectives) there is a 0.142 (14.2%) chance of finding one or more defectives in the sample and therefore making an unnecessary adjustment to the process. Yet with the furnace operating badly there is a 0.512 chance of finding zero defectives in the sample and consequently making no adjustment to a faulty process.

We have so far considered a rather impractical situation in which the sample size was only 3. Practical situations will call for far larger sample sizes which clearly will be beyond using a probability tree for the calculations. Three alternative ways of obtaining binomial probabilities are:

(a) using a table of probabilities such as Table A.10;
(b) using a computer program;

Table 11.1   Binomial distribution for 10% defectives and a sample size of 3

No. of defectives	0	1	2	3	Total
Probability of obtaining specified number of defectives	0.729	0.243	0.027	0.001	1.000

Table 11.2   Binomial probabilities for a sample size of 3

No. of defectives		0	1	2	3	Total
Probability of obtaining	$\pi = 5\%$	0.857	0.135	0.007	0.0001	1.000
specified number	$\pi = 10\%$	0.729	0.243	0.027	0.001	1.000
of defectives	$\pi = 20\%$	0.512	0.384	0.096	0.008	1.000

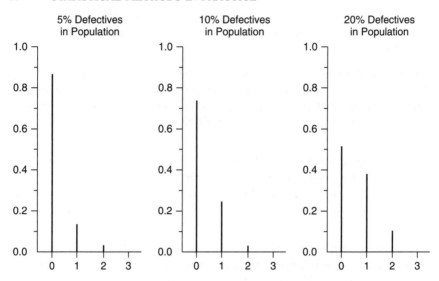

Figure 11.2   Binomial probabilities with a sample size of 3

(c) using a calculator with the formula

$$\text{Probability} = \frac{n!}{r!(n-r)!}\pi^r(1-\pi)^{n-r}$$

where r = number of defectives in the sample,
n = sample size,
π = proportion of defectives in the population.
$n! = n \times (n-1) \times \ldots 3 \times 2 \times 1 (0! = 1)$.
For example, with $r = 1, n = 3$ and $\pi = 0.1$

$$\text{Probability} = \frac{3!}{1!(3-1)!}(0.1)^1(1.0-0.1)^{3-1}$$

$$= \frac{3 \times 2 \times 1}{1 \times (2 \times 1)}0.1^1 0.9^2 = 0.243$$

The value of 0.243 is of course identical to that obtained using the probability tree in Figure 11.1.

Let us now look at a more realistic control scheme in which the process engineer takes a sample of 20 insulators and only makes an adjustment to the process if three or more defectives are present.

The probabilities listed in Table 11.3 have been abstracted from Table A.10 for a sample size of 20. They do not, however, include the full range of numbers of defectives since the majority of them have extremely small probabilities.

Table 11.3   Binomial probabilities for a sample size of 20

No. of Defectives		0	1	2	3 or more	Total
Probability of obtaining	$\pi = 5\%$	0.359	0.377	0.189	0.075	1.000
specified no. of	$\pi = 10\%$	0.122	0.270	0.285	0.323	1.000
defectives	$\pi = 20\%$	0.012	0.058	0.137	0.794	1.000

Therefore for large numbers of defectives the probabilities have been combined under the heading '3 or more'. Since total probability sums to 1.0, the probability of 3 or more can easily be found by subtracting the probabilities for 0, 1 and 2 defectives from 1.0. For example, with $\pi = 5\%$, the probability of three or more defectives is

$$= 1.0 - \{\text{Probability of 0 or 1 or 2 defectives}\}$$

$$= 1.0 - \{0.359 + 0.377 + 0.189\}$$

$$= 1.0 - 0.925 = 0.075$$

Examination of Table 11.3 shows that the process engineer has achieved a better control scheme. If the furnace is operating well there is only a 0.075 chance of making an unnecessary adjustment, while if it is operating badly there is a 0.793 chance of rightly deciding to adjust the process.

An exact binomial test

The quality of an insulator is mainly determined by whether it is hazed or not, but any difference in appearance may cause concern to a customer who might assume – probably wrongly – that the difference indicates a deterioration in quality due to a change in the process. Therefore, Nocharge inspects each batch to ensure that the insulators look similar to the previous batch. (This is considered reasonable since a customer is likely to compare insulators from adjacent batches but not from ones produced several months apart.)

The inspection procedure uses the triangle test, which is well known in the sensory analysis of food. The order of presentation is given in Table 11.4.

Table 11.4   Presentation of insulators

Group	1	2	3	4	5	6	7	8	9	10	11	12
Left	N	P	P	N	N	P	N	P	P	N	N	P
Centre	P	N	P	N	P	N	P	N	P	N	P	N
Right	P	P	N	P	N	N	P	P	N	P	N	N

N = new batch; P = previous batch

The inspector is presented with 12 groups each containing three insulators and he is asked to choose the insulator which looks different from the other two. Clearly if there is a marked difference between the batches in **any** characteristic he should be able to choose the 'odd one out' correctly. If, however, there is no difference the probability of choosing correctly is only one in three. Let us suppose an inspector chooses 6 out of 12 correctly. Does this indicate a difference between the batches or is it just due to chance?

Null hypothesis	$\pi = \frac{1}{3}$. The inspector cannot determine any difference between the previous batch and the new batch.
Alternative hypothesis	$\pi > \frac{1}{3}$. The inspector can determine a difference between the batches.
Test value	6 is the number of correct determinations.
Table values	From Table A.13 ($\pi = \frac{1}{3}$) with 12 trials: $7\frac{1}{2}$ at the 5% significance level; $8\frac{1}{2}$ at the 1% significance level.
Decision	Since the test value is less than the table value we cannot reject the null hypothesis.
Conclusion	We cannot conclude that there is a difference in appearance between insulators from the two batches.

Thus Nocharge are quite prepared to accept that no major change has occurred to the visual appearance of the insulators.

This form of test is necessary whenever we cannot define clearly the difference we are seeking. Although an inspector might find the batches to be different, he did not have to define the characteristic which makes them different. When we can state the characteristic – for example, 'which is more hazed?' – then it is more efficient to use pairs of samples and carry out a significance test with a null hypothesis of $\pi = \frac{1}{2}$ and a table value from Table A.13 ($\pi = \frac{1}{2}$).

A quality assurance example

Unifoods buy screw-top glass bottles from Glasstight in large batches of approximately 50 000 bottles. Unifoods have developed a manual test to show whether bottles are airtight when fitted with a top.

Unifoods and Glasstight have jointly agreed to define a sampling inspection scheme.

The criterion that Unifoods lay down is that no batch should be released which has more than 2% defective bottles (that is, not airtight).

For Glasstight the criterion is that no batch should be rejected if it has less than 0.5% defective bottles.

Unifoods and Glasstight decide to look at two possible schemes.

Scheme A: Select **100** bottles at random.
Test, and **reject** the batch if there is **one failure or more.**

Scheme B: Select **100** bottles at random.
Test, and if **two or more fail, reject** the batch.

The probability of whether a batch is accepted or rejected will depend on the percent defectives in the **batch** of 50 000. These probabilities can be obtained from Table A.10 or by using *Crunch*.

The output from *Crunch* is shown below.

Single Sampling Scheme		
Percent	A	B
0.00	1.000	1.000
0.25	0.779	0.974
0.50	0.606	0.910
0.75	0.471	0.827
1.00	0.366	0.736
1.25	0.284	0.644
1.50	0.221	0.557
1.75	0.171	0.476
2.00	0.133	0.403
2.25	0.103	0.339
2.50	0.080	0.283
2.75	0.062	0.235
3.00	0.048	0.195
3.25	0.037	0.160
3.50	0.028	0.131
3.75	0.022	0.107
4.00	0.017	0.087
4.25	0.013	0.071
4.50	0.010	0.057
4.75	0.008	0.046
5.00	0.006	0.037

There are two curves on the graph, one for each scheme.

Each scheme is described by the sample size and the acceptance number, which is the maximum number of failures in the sample for which a batch will be accepted. For Scheme A the acceptance number is 0 and for Scheme B it is 1.

The curves are known as operating characteristic (OC) curves. For any percentage of defectives in the batch it shows the probability of a batch being accepted.

Scheme A gives a very flat curve, which is not ideal.

Scheme B is flat at the start and flat at the end but changes gradient slowly in the middle. This is a better shape but, ideally, the gradient should change very quickly between the 0.5% defectives that Glasstight desires and the 2% that Unifoods can tolerate.

If we examine the table of probabilities all is not good.

With Scheme A there is only a 0.606 probability of accepting a batch with 0.5% defectives which is totally unacceptable to Glasstight as it means that there is a 0.394 probability of such a batch being rejected. There is, however, only a probability of 0.133 that a batch with 2% defectives will be accepted, which would suit Unifoods.

With Scheme B there is a 0.910 probability of accepting a batch with 0.5% defectives, which is suitable for Glasstight, but a probability of 0.403 of accepting a batch with 2% defectives, which is completely unacceptable for Unifoods.

The only way of redressing the balance in the schemes is to increase the sample size. They consider a third scheme.

Scheme C: Take a sample of 340 bottles.
                 Reject the batch if there are 4 or more failures in the sample.

The output from *Crunch* shows this scheme compared with Scheme A.

Single Sampling Scheme	A	C
Sample Size	100	340
Acceptance No.	0	3

Upper limit for graph(%)   5

Percent Defectives

Single Sampling Scheme		
Percent	A	C
0.00	1.000	1.000
0.25	0.779	0.989
0.50	0.606	0.907
0.75	0.471	0.747
1.00	0.366	0.558
1.25	0.284	0.385
1.50	0.221	0.249
1.75	0.171	0.153
2.00	0.133	0.091
2.25	0.103	0.052
2.50	0.080	0.029
2.75	0.062	0.016
3.00	0.048	0.008
3.25	0.037	0.004
3.50	0.028	0.002
3.75	0.022	0.001
4.00	0.017	0.001
4.25	0.013	0.000
4.50	0.010	0.000
4.75	0.008	0.000
5.00	0.006	0.000

The acceptance number for Scheme C is 3.

We can see that Scheme C has higher probabilities than Scheme A at the start and lower at the end. Examining the probabilities in the table gives the following conclusions:

There is a 0.907 probability of accepting a batch with 0.5% defectives and a 0.091 probability of accepting a batch with 2% defectives. Unifoods and Glasstight are delighted. There are similar risks of making wrong decisions for both their criteria of good and bad batches. Glasstight know that as long as the quality (percentage of defectives in the batch) is below 0.5%, few batches will be failed; Unifoods know that if a bad batch is produced (2% or more defectives in the batch) the sampling procedure will isolate it.

What is the effect of the batch size?

The sample of 340 was taken from a batch of 50 000 bottles, but the size of the batch did not feature in our calculations. This is correct, providing that the batch size is far larger than the sample size, for the simple reason that the information we glean about the batch is taken from the sample. The larger the sample, the greater the information, irrespective of the batch size.

There is, however, a difficulty with large batch size: obtaining a representative sample. Thus it is generally accepted that the sample size should increase, but not proportionally, as the batch size increases in order to make it feasible to obtain a representative sample.

Problems

11.1 Super Mark, a chain of European supermarkets, buys 500 g jars of mayonnaise from Sandringham Sauces in deliveries of 6000. A quality inspector takes a sample of 20 jars at random from a delivery and checks them for leaking seals. The delivery is rejected if he finds any defective seals.

(a) Which statistical distribution can be applied to the number of defective seals in a sample?

(b) If 1% of the jars in a delivery have defective seals, what is the probability of one (or more) defectives being found in a sample of 20?

(c) What is the probability of a delivery which contains 1% defective being rejected?

(d) Which of the sample sizes in Table A.10 would be needed so that the probability of rejecting a batch containing 1% defectives would be at least 0.50 on the basis of at least one defective being found in the sample?

11.2 Cocoanuts plc purchase many consignments of cocoa beans to make their chocolate products. However, being a natural material, there is considerable variability in quality and therefore Cocoanuts always subject each consignment to an inspection scheme.

This consists of taking a sample of beans, cutting each bean in half and classing a bean as 'defective' if the inside is mouldy, or 'slaty' (greyish tinge) in colour (showing it has not matured).

Any consignment with 5% or more defectives is classed as second grade and used for inferior products.

Cocoanuts wish to devise a suitable inspection scheme which will accept consignments if the defective rate is 3% or less and reject consignments if it is 5% or more.

Initially they believe a sample size of 100 beans and an acceptance number of 3 or 4 will produce a reasonable scheme.

(a) Using *Crunch*, obtain the operating characteristic curves for the two schemes. From the graph and the table of probabilities obtain probabilities of accepting a good batch (3% defectives) and a bad batch (5% defectives). Decide whether either scheme is satisfactory.

(b) By changing the sample size and acceptance number in *Crunch*, obtain a scheme in which the probabilities of rejecting a good batch and accepting a bad batch are both approximately 0.1.

12

The Poisson distribution

Introduction

The Poisson distribution is another distribution describing discrete data. Examples are:

- counts of bacteria per plate in homogeneous liquids;
- number of oversize particles in a perfectly distributed medium;
- number of breakdowns caused by many different faults.

We shall use an example from the textile industry to illustrate where and when to use the Poisson distribution.

Fitting a Poisson distribution

Accord Ltd are weavers and dyers of fabrics for the new European Navy who have the highest possible standards; they are only prepared to accept fabrics with one or two minor defects. As part of the agreed quality programme, Accord inspect every tenth roll and if the number of defects exceeds 5 (remember these are minor defects), they subject the preceding rolls to full inspection. Any rolls failing full inspection are sold as sub-standard to a well-known market trader in Knaresborough.

The managing director of Accord has become concerned about the losses due to sub-standard rolls and has asked his textile technologist, William Pilling, to investigate. Pilling immediately recognises the wealth of information available in old inspection records and decides to concentrate initially on records from the previous six months (July to December). He carries out an analysis of defects and

Statistical Methods in Practice: for Scientists and Technologists   R. Boddy, G. Smith,
© 2009 John Wiley & Sons, Ltd

Table 12.1   Number of dye-flecks in each roll

Roll	No.	Roll	No.	Roll	No.	Roll	No.	Roll	No.	Roll	No.	Roll	No.	Roll	No.
1	0	11	2	21	1	31	1	41	1	51	1	61	3	71	0
2	3	12	2	22	1	32	1	42	0	52	2	62	3	72	2
3	3	13	2	23	1	33	2	43	0	53	4	63	4	73	2
4	3	14	0	24	1	34	1	44	4	54	1	64	1	74	3
5	2	15	2	25	0	35	1	45	0	55	1	65	1	75	3
6	2	16	3	26	0	36	3	46	0	56	1	66	0	76	3
7	1	17	1	27	0	37	0	47	2	57	0	67	0	77	0
8	0	18	5	28	2	38	0	48	0	58	0	68	0	78	1
9	1	19	1	29	1	39	2	49	2	59	4	69	1	79	1
10	1	20	1	30	1	40	3	50	2	60	1	70	1	80	0

Table 12.2   Frequency distribution of dye-flecks

No. of dye-flecks	0	1	2	3	4	5
No. of rolls with stated no. of dye-flecks	21	28	15	11	4	1

finds that 96% of defects are caused by dye-flecks (small lengths of undrawn yarn which dye far darker than the rest of the fabric).

The recorded numbers of dye-flecks found in the rolls inspected during that period are shown in Table 12.1. Pilling computes a frequency distribution of these defects, which is shown in Table 12.2.

Are the defects random? The Poisson distribution

The Poisson distribution occurs when an event occurs at random in either time or space and we record the number of events in a fixed interval. In this context the 'event' is a dye-fleck and the 'interval' is a roll of fabric. Thus if a dye-fleck occurs at random, the numbers per roll will follow a Poisson distribution. A Poisson distribution is defined by one parameter, its population mean (μ). It depends solely upon the mean, as shown in Figure 12.1.

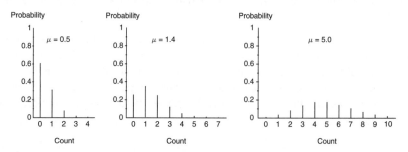

Figure 12.1   Poisson distributions

The probabilities in Figure 12.1 have been obtained from Table A.11.

With a population mean (μ) of 0.5 the distribution is highly skewed whereas with a mean of 5.0 or above, it resembles a normal distribution, a fact which is often used to simplify calculations. Another important property of the Poisson distribution is

$$\text{Standard deviation} = \sqrt{\mu}$$

A Poisson distribution also has the characteristic that no matter how recently an event has occurred, the probability of another event occurring in any time-period or distance is unchanged.

Let us now fit a Poisson distribution to the dye-fleck data.

We first estimate the mean number of dye-flecks per roll of fabric.

$$\text{Mean} = \frac{(21 \times 0) + (28 \times 1) + (15 \times 2) + (11 \times 3) + (4 \times 4) + (1 \times 5)}{21 + 28 + 15 + 11 + 4 + 1}$$

$$= \frac{112}{80} = 1.40 \text{ dye-flecks per roll}$$

Turning to Table A.11, we obtain the probability for a stated number of defects. For example, with a mean of 1.40:

Probability of 2 dye-flecks   = 0.2417
Expected frequency for 2 dye-flecks   = 0.2417 × (total no. of rolls)
                                          = 0.2417 × 80
                                          = 19.3

Thus, if dye-flecks followed a Poisson distribution with a mean of 1.40 per roll, we could have expected to have found 19.3 rolls with two faults. In actual fact we found 15. We can proceed to compute a table of expected frequencies as shown in Table 12.3.

The data from Table 12.3 are plotted as a bar chart in Figure 12.2. This shows excellent agreement between observed and expected frequencies. It is therefore reasonable to assume that dye-flecks follow a Poisson distribution and they are therefore random. This is not perhaps to the liking of William Pilling. Randomness implies many causes and reducing the frequency will be difficult. Non-randomness often implies a single cause and if this cause can be found and corrected the number of dye-flecks would be greatly reduced.

If the mean was not one of those included in Table A.11 we would need to use *Crunch* or the following formula for Poisson probabilities:

$$\text{Probability of } r \text{ defects} = \frac{\mu^r e^{-u}}{r!}$$

where r is the number of defects in a defined interval, μ is the mean number of defects, e is the exponential constant, equal to approximately 2.718, and $r!$ is 'r factorial' $= r \times (r - 1) \times \ldots \times 3 \times 2 \times 1$ (note that $0! = 1$).

Table 12.3   Expected frequencies for dye-flecks using the Poisson distribution

No. of dye-flecks per roll	Observed frequency	Poisson probability ($\mu = 1.40$)	Expected frequency
0	21	0.2466	19.7
1	28	0.3452	27.6
2	15	0.2417	19.3
3	11	0.1128	9.0
4	4	0.0395	3.2
5	1	0.0111	0.9
6 and above	0	0.0031	0.3
Total	80	1	80.0

Frequency

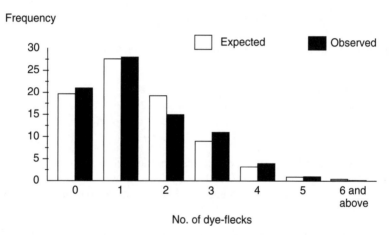

Figure 12.2   Comparison of expected and observed frequencies

Poisson dispersion test

A simple test of random dispersion of defects is based on the fact that **in a Poisson distribution the standard deviation is the square root of the mean** or, equivalently, **the variance equals the mean**. For a normal distribution, it will be recalled that both the mean and the standard deviation need to be determined. In a Poisson distribution, however, only the mean is required.

The sample mean number of dye-flecks was 1.40.

If Pilling's data came from a Poisson distribution, then the standard deviation would be near to the square root of 1.40, which is 1.18.

From the data in Table 12.1, Pilling determined that the sample standard deviation was 1.23, which is higher than the standard deviation of 1.18 which we would expect if the numbers of defects had a Poisson distribution. Did this occur by chance?

We can carry out a 'dispersion' test to see whether the data could have come from a Poisson distribution.

Null hypothesis   The dye-flecks are randomly distributed (their numbers follow a Poisson distribution).

Alternative hypothesis   The dye-flecks are not randomly distributed (their numbers do not follow a Poisson distribution).

Test value

$$\text{Dispersion index} = \frac{SD^2}{\text{Mean}}$$

$$= \frac{(1.23)^2}{1.40} = 1.08$$

Table values   From Table A.8 (Poisson dispersion test, two-sided) with number of degrees of freedom equal to (number of rolls − 1) = 79:

0.71–1.34 at the 5% significance level.

Decision   The test value lies inside the range of the table values. We cannot reject the null hypothesis.

Conclusion   We conclude that the dye-flecks are randomly distributed.

Confidence intervals for a Poisson count

The observed mean ($\bar{x}$) which Pilling obtained from his data was 1.40 dye-flecks per roll.

A 95% confidence interval for the true mean is given by:

$$\bar{x} \pm 1.96\sqrt{\frac{\bar{x}}{n}}$$

where $\bar{x}$ is the observed mean, and n is the number of rolls.

The confidence interval is therefore:

$$1.40 \pm 1.96\sqrt{\frac{1.40}{80}} = 1.40 \pm 0.26$$

$$= 1.14 \text{ to } 1.66$$

Pilling can be 95% confident that the true mean number of dye-flecks is between 1.14 and 1.66 per roll.

A significance test for two Poisson counts

Undaunted by the conclusions regarding randomness, William Pilling continues his search for source of dye-fleck defects and stumbles upon data which shows that in the previous six months (January to June) there were only 1.13 dye-flecks per roll, the actual number being 113 in 100 rolls. However, caution warns him that, before looking for the cause, he should carry out a significance test to decide if there is a real difference between the two periods.

Null hypothesis For all rolls produced in the two periods the dye-fleck rate for January to June was the same as for July to December.

Alternative hypothesis The dye-fleck rates were different.

Test value

$$= \frac{\left| \dfrac{\text{Count}_A}{n_A} - \dfrac{\text{Count}_B}{n_B} \right|}{\sqrt{\dfrac{\text{Count}_A + \text{Count}_B}{n_A \times n_B}}}$$

where Count_A, Count_B are the total number of dye-flecks in each period and n_A, n_B are the number of rolls.

For the period January to June, $\text{Count}_A = 113$, $n_A = 100$
For the period July to December, $\text{Count}_B = 112$, $n_B = 80$

The test value becomes:

$$\frac{\left| \dfrac{113}{100} - \dfrac{112}{80} \right|}{\sqrt{\dfrac{113 + 112}{100 \times 80}}} = \frac{|1.13 - 1.40|}{0.168} = 1.61$$

Table values Using Table A.1 (normal distribution, two-tailed) gives:
1.96 at the 5% significance level;
2.58 at the 1% significance level.

Decision We cannot reject the null hypothesis at the 5% level.

Conclusion There is not enough evidence to indicate that the defective rates in the two periods are different.

Reluctantly William Pilling has to accept the 'not significant' verdict and that he has again drawn a blank in his search for the cause of the defects.

How many black specks are in the batch?

Silicaceous Ltd produce a powder for toothpaste which gives a 'mouth-fresh' feel to the toothpaste. Unfortunately, Silicaceous have a production problem which has proved unsolvable – that of the occasional small black speck in the powder. Their customer, Feelfresh Toothpaste Ltd, is aware of the problem and is not concerned, providing the number of specks is less than one per 10 tubes of toothpaste.

Sicilaceous take a sample of 200 g from a batch of 20 kg and examine it for black specks. They find two specks in the sample.

What can we say about the amount in the batch?

Clearly they have taken a sample and the results are subject to sampling variation. We should obtain a confidence interval for the batch and since Feelfresh is only concerned with high numbers of specks we shall use a **one-sided** confidence interval, giving us an **upper** confidence limit.

We can obtain such a limit by looking at Table A.16. The 'observed count' was 2. The table gives a 95% upper limit of 6.3 for a count of 2.

Thus we can say that we are 95% certain that the number of black specks in the batch averages less than 6.3 in 200 g of powder.

How does this compare with Feelfresh's specification?

Ten tubes of toothpaste use 30 g of powder.

Therefore for 30 g the upper limit will be:

$$\frac{6.3 \times 30}{200} = 0.95 \text{ specks.}$$

This is just under the specification! The batch is passed!

Silicaceous are concerned. They have just escaped with that batch, but if there had been 3 or more specks in the 200 g sample they would have had an upper limit, as given in Table A.16, of 7.7 in 200 g, or 1.16 in 30 g and the batch would have been failed. They feel than such a scheme is rather tight.

They decide to look at a 300 g sample.

With 300 g Feelfresh's specification is less than 10 specks in a batch.

Examining Table A.16 we see that this will be achieved if there are 4 or fewer specks in the sample. The upper confidence limit with a count of 4 is 9.1.

Silicaceous are better pleased with the scheme, but the success of it can only be judged when they have tested a number of batches. Production will need to average considerably less than 4 specks per 300 g, otherwise a number of failures will occur.

How many pathogens are there in the batch?

The presence of salmonella in milk powder can have serious outcomes and clearly the specification is that none should be present. Drifoods Limited produce 2000 kg batches of milk powder in a spray drier. From this they take 80 samples of 25 g at regular intervals to obtain a representative sample of the batch. The tests are all negative.

What can be said about the batch of 2000 kg?

Table A.16 gives a 95% upper confidence limit of 3.0 with a count of zero. Our sample size was $80 \times 25\,g = 2\,kg$. Thus we are 95% certain that the density of salmonella organisms is less than 3.0 per 2 kg.

Applying this to the batch of 2000 kg gives an 95% upper limit of 3000 salmonella organisms.

Drifoods are shocked by the result. Clearly they have done a large amount of testing but have achieved little quality assurance. Increasing the testing is not a serious option. If they increase the amount of testing tenfold they would still be facing an upper limit of 300 organisms. Thus this test is only a rudimentary check on the presence of salmonella. Quality assurance must be concerned with ensuring that salmonella does not pervade the process, rather than testing to see if it is present – good manufacturing practice must be in place.

Problems

12.1 2TfruT Tarts are well known for their high-quality tarts but they have recently been running into problems – some 10% of tarts are being classed as defective; that is, they have too few strawberries. Usually if there is no more than one strawberry in a tart, it is classed as defective. The managing director of 2TfruT is annoyed. He has specified that each vat of strawberry and jelly mix should contain enough strawberries to average four per tart. He instructs the works manager, Erdberry, to investigate whether his instructions are being followed.

Erdberry quickly counts the strawberries in 60 tarts, and obtains a total of 228.

(a) Calculate a confidence interval for the mean and hence conclude whether or not the managing director's instructions are being followed. What assumptions have been made?

(b) The managing director concludes that if the average is correct, the mixing must be poor, so he doubles the length of mixing time – without effect. He therefore instructs Erdberry to find out why the mixing is so poor despite doubling the time. Erdberry now carries out a more detailed examination and obtains a frequency distribution for 90 tarts.

No. of strawberries	No. of tarts
0	2
1	7
2	11
3	16
4	21
5	13
6	11
7	5
8	3
9	1

Compute expected frequencies using the Poisson distribution and draw a multiple bar chart of observed and expected frequencies.

(c) Does the Poisson distribution provide a good fit? What should Erdberry tell his managing director?

(d) What should the average number of strawberries be so that the percentage defectives is only 1% (use *Crunch*).

12.2 (a) Precision Printers are concerned with the number of paper jams on their equipment. Over a 30-day period they had 93 jams. Assuming these occurred at random, calculate a 95% confidence interval for the mean number of jams per day.

(b) To reduce the problem, Precision Printers decided to buy higher-quality paper. In the first 20 days they had 41 jams. Carry out a test to determine whether there has been a significant improvement.

13

The chi-squared test for contingency tables

Introduction

This chapter introduces the use of the chi-squared test for the comparison of two or more sample percentages. It may be used, for example, to compare percentage preferences for a product by different groups, or to compare gradings of products into acceptable/not acceptable.

Two-sample test for percentages

Manscent are a company specialising in cosmetics for men. They are currently investigating who purchases their products and seek to target the products to the appropriate market, as they realise some are purchased by men for their own use and some are purchased by women for their men. Two aftershaves, 'David' and 'Goliath', which have different scents, were presented to 80 men and 100 women, who were asked to state their preference. The numbers of preferences are shown in Table 13.1.

We can see from Table 13.1 that of the respondents, 47 out of 80, or 59% of the men preferred Goliath, while 66 out of 100, or 66% of the women preferred David. Does that indicate a real difference between the preferences of the two groups of customers, or was the difference between the preferences due to chance? A suitable significance test proceeds as follows:

Statistical Methods in Practice: for Scientists and Technologists   R. Boddy, G. Smith,
© 2009 John Wiley & Sons, Ltd

Table 13.1   Number of preferences for each aftershave

Aftershave Respondents	David	Goliath	Total
Men	33 (41%)	47 (59%)	80
Women	66 (66%)	34 (34%)	100
Total	99 (55%)	81 (45%)	180

Null hypothesis

The population percentage preference for Goliath is the same in both groups.

Alternative hypothesis

The population percentage preference for Goliath is different in the two groups.

Test value

$$= \frac{|P_1 - P_2|}{\sqrt{\bar{P}(100 - \bar{P})\left(\dfrac{1}{n_1} + \dfrac{1}{n_2}\right)}}$$

where P_1, P_2 are the observed percentages preferring 'Goliath' in each of the groups (59 and 34),

n_1, n_2 are the total numbers of respondents in each group (80 and 100),

and $\bar{P}$ is the percentage of all respondents preferring 'Goliath' (45)

The test value is therefore:

$$\frac{|59 - 34|}{\sqrt{45(100 - 45)\left(\dfrac{1}{80} + \dfrac{1}{100}\right)}} = \frac{25}{7.46} = 3.32$$

Table value

From Table A.1 (two-tailed):
1.96 at the 5% significance level,
2.58 at the 1% significance level.

Decision

Reject the null hypothesis at the 1% significance level.

Conclusion

The percentage preference for Goliath differs between men and women.

Crunch output

Aftershave Respondents	David		Goliath		Total
Men	33	41%	47	59%	80
	33%		58%		44%
Women	66	66%	34	34%	100
	67%		42%		56%
Total	99	55%	81	45%	180

Test Value	3.32
Significance Level	5%
Table value	1.96
Decision	Significant

The preferences are shown, as in the chapter, as numbers and row percentages of men and of women. Notice the column percentages are also shown, to allow for the data being entered in either way, in rows or in columns.

Comparing several percentages

Nocharge plc are major producers of electrical insulators. A fault which will affect performance is 'hazing' of the surface. Every insulator is inspected as it leaves the furnace. The management is concerned that the number of hazed insulators, currently running at 8% of production, is too high, and also that there may be some difference in quality between the four shifts. The numbers of hazed insulators have been separated into the different shifts, as shown in Table 13.2. The sample percentages show differences between the shifts, but could they just be due to chance?

To compare the percentages of hazed insulators from the four shifts, we use a **chi-squared** significance test.

First, we need to recognise that each insulator is either 'hazed' or 'clear' and we need to rearrange Table 13.2 to show the numbers of insulators in each of those categories.

Table 13.2   Numbers and percentages of hazed insulators from each shift

Shift	Red	Blue	Green	Black	Total
No. of insulators sampled	200	210	330	320	1060
No. hazed	9	7	11	26	53
Percentage hazed	4.5%	3.3%	3.3%	8.1%	5.0%

Table 13.3   Numbers and hazed and clear insulators from each shift

Shift	Red	Blue	Green	Black	Total
No. hazed	9	7	11	26	53
No. clear	191	203	319	294	1007
Total	200	210	330	320	1060

Table 13.4   Observed and expected frequencies

	Red		Blue	Green	Black	Total
Hazed	9 (Observed)		7	11	26	53
	(Expected)	10.0	10.5	16.5	16.0	
Clear	191		203	319	294	1007
		190.0	199.5	313.5	304.0	
Total	200		210	330	320	1060

The test compares the frequencies in Table 13.3 with those that would be expected if the four sample percentages had all been equal.

Of the 1060 insulators, 53 or 5.0% were hazed.

If the same percentage of insulators were hazed in each shift, we should expect the number of hazed insulators from the Red shift to be 5.0% of 200 = 10.

A convenient formula to calculate the expected frequency in each category is as follows:

$$\text{Expected frequency} = \frac{\text{Row total} \times \text{Column total}}{\text{Overall total}}$$

For example, the expected frequency in the first row and first column is:

$$\frac{53 \times 200}{1060} = 10.0$$

The full set of expected frequencies is shown in Table 13.4. Notice that the row and column totals in Table 13.4 are unchanged from those in Table 13.3.

We can now proceed to a significance test.

Null hypothesis The population percentage of hazed insulators is the same for all four shifts.

Alternative hypothesis The population percentage of hazed insulators is not the same for all four shifts.

Test value
$$= \sum \frac{(\text{Observed frequency} - \text{Expected frequency})^2}{\text{Expected frequency}}$$

$$= \frac{(9 - 10.0)^2}{10.0} + \frac{(7 - 10.5)^2}{10.5} + \frac{(11 - 16.5)^2}{16.5}$$
$$+ \frac{(26 - 16.0)^2}{16.0}$$

$$= + \frac{(191 - 190.0)^2}{190.0} + \frac{(203 - 199.5)^2}{199.5}$$
$$+ \frac{(319 - 313.5)^2}{313.5} + \frac{(294 - 304.0)^2}{304.0}$$
$$= 9.84$$

Table value

From Table A.8 (one-sided for contingency tables), for degrees of freedom
= (no. of rows − 1) × (no. of columns − 1)
= (2 − 1)(4 − 1) = 3 :
7.81 at the 5% significance level,
11.34 at the 1% significance level.

Decision

Reject the null hypothesis at the 5% level.

Conclusion

The percentage of hazed insulators is not the same for all four shifts.

Where are the differences?

Now that we have concluded that the shifts are not giving the same quality of insulators, we need to determine where the significant differences lie.

We can determine a least significant difference (LSD). If the two percentages differ by more than this amount, they are significantly different; if they do not, they are not significantly different.

A 5% LSD between two percentages is given by:

$$5\% \text{ LSD} = 1.96 \sqrt{\bar{P}(100 - \bar{P}) \left(\frac{1}{n_1} + \frac{1}{n_2} \right)}$$

where $\bar{P}$ is the total percentage of hazed insulators, and n_1, n_2 are the sample sizes for the two percentages being compared.

For each comparison, we should use the actual two samples sizes, but if the samples sizes are not very different we can use the average sample size (265 per shift) for

n_1 and n_2, and obtain:

$$5\% \text{ LSD} = 1.96\sqrt{5.0 \times (100 - 5.0) \left(\frac{1}{265} + \frac{1}{265}\right)} = 3.7\%$$

The percentages for the four shifts were: Red 4.5%, Blue 3.3%, Green 3.3%, Black 8.1%.

Thus the problem in the defective rates was Black shift, whose percentage of 8.1% was significantly higher than that of Blue and Green and marginally worse than that of Red. The other three shifts had no significant differences among them.

The comparisons can be shown graphically by listing the percentages in order of their magnitude and bracketing together those which are **not** significantly different from each other:

Black	8.1%
Red	4.5%
Blue	3.3%
Green	3.3%

Crunch output

Grade	Shift	Red		Blue		Green		Black			Row Total
Hazed		9	10.0	7	10.5	11	16.5	26	16.0		53
Clear		191	190.0	203	199.5	319	313.5	294	304.0		1007
											0
											0
											0
Column Total		200		210		330		320		0	1060

Test Value	9.84
df	3
Significance Level	5%

Table Value	7.81

Decision	Significant

Least Significant Difference
between percentages
(data in 1st two rows only

n_1	265
n_2	265
sig. level	5%

5% LSD	3.71%

Assumptions

The chi-squared test is based on an approximation which works well providing the sample size is large enough. In carrying out the chi-squared test we should be aware that a condition for the test is that the **expected frequencies are at least five**. In this example the expected frequencies were 190 or more.

The chi-squared test can be applied to any contingency table in which frequencies are classified by both row and column.

The calculation of least significance difference applies only to the situation in the example in the chapter, where data are entered in the first two rows relating to two possible outcomes, for example an insulator being 'hazed' or 'clear' and it is desired to compare the percentage hazed between different situations – for example, shifts.

Problems

13.1 Nocharge plc are major producers of electrical insulators. A fault which will affect performance is 'hazing' of the surface. Every insulator is inspected as it leaves the furnace. The management is concerned that the number of hazed insulators, currently running at 8% of production, is too high. A modification has been made to the process and the Production Manager wishes to know whether the percentage of hazed insulators has changed.
The results of inspections before and after the modification were as follows:

	Before	After	Total
No. of insulators sampled	500	300	800
No. hazed	40	12	52

Has the percentage of hazed insulators changed significantly after the modification?

13.2 Cocoanuts plc purchase many consignments for cocoa beans to make their chocolate products. They have an acceptance sampling scheme which works well, provided the consignment is homogeneous. However, they believe a certain grower is mixing second-grade beans within the rest of a high-quality consignment. They decide to investigate by taking four samples at different levels from within the consignment. They cut the beans in half and then class them as: mouldy; slaty (grey in colour, showing the beans are not mature); or first class (with none of the above defects).
The results are as follows:

Level Grade	Top	Upper middle	Lower middle	Bottom
Mouldy	11	9	16	13
Slaty	5	10	8	19
First class	473	481	461	417

(a) Carry out a significance test, obtaining the test value from *Crunch*. Is there any evidence that the consignment is not homogeneous?

(b) From the data it appears that the 'slaty' category varies between levels. Using slaty and first-class categories, obtain a least significant difference to decide which levels are significantly different.

14

Non-parametric statistics

Introduction

The most commonly used statistical tests are based on the normal distribution. The sample mean and standard deviation are the appropriate measures to estimate the centre and spread of a normal distribution but may not be appropriate for a population which has a different shape of distribution. Table values for the tests such as the t-test, the F-test and the correlation coefficient all assume a normal distribution of data or residuals. Any other distribution would require different table values or a modification of the test.

There are situations where we may know that data will not be normally distributed but we may not be able to say what type of distribution will apply. A family of statistical methods which make no assumptions about the distribution of populations is useful in these situations. These methods are called **non-parametric** or **distribution-free**.

Descriptive statistics

In describing data which may come from a population with any shape of distribution, the most generally useful measure to describe the centre of the population is the **median**.

The median is a good measure because, irrespective of the shape of the distribution, half the population is below the median and half is above the median.

It is estimated by the **sample median**.

As part of a regular monitoring procedure, the sulfur dioxide emissions from Dalechem's plant are recorded daily. The figures for one month in ppm are shown in Table 14.1.

Statistical Methods in Practice: for Scientists and Technologists R. Boddy, G. Smith,
© 2009 John Wiley & Sons, Ltd

Table 14.1   Daily sulfur dioxide emissions

Day	1	2	3	4	5	6	7	8	9	10	11	12	13	14
Level	15	17	12	23	35	19	8	45	22	13	17	19	16	8

Day	15	16	17	18	19	20	21	22	23	24	25
Level	135	42	26	16	9	13	17	15	24	20	27

Table 14.2   Data sorted in order

Order	1	2	3	4	5	6	7	8	9	10	11	12	13	14
Day	7	14	19	3	10	20	1	22	13	18	2	11	21	6
Level	8	8	9	12	13	13	15	15	16	16	17	17	17	19

Order	15	16	17	18	19	20	21	22	23	24	25
Day	12	24	9	4	23	17	25	5	16	8	15
Level	19	20	22	23	24	26	27	35	42	45	135

The sample median is the middle observation when the data have been sorted in order. The sorted values are shown in Table 14.2.

As there are 25 observations, an odd number, the sample median is taken to be the 13th value, 17 ppm.

Two other useful summary statistics are the **lower and upper quartiles**, which, along with the median, divide the population into parts of equal numbers.

The lower quartile, or 25th percentile, is the value which has 25% of the population less than it. It is estimated by the value whose order is $(n + 1)/4 = 26/4 = 6.5$. It is half-way between the 6th and 7th ordered observations, $(13 + 15)/2 = 14$ ppm.

The upper quartile, or 75th percentile, is estimated by the value whose order is $3(n + 1)/4 = 78/4 = 19.5$. It is half-way between the 19th and 20th ordered observations, $(24 + 26)/2 = 25$ ppm.

The **interquartile range**, or difference between the upper and lower quartiles, is often used as a non-parametric measure of spread.

Interquartile range $= 25 - 14 = 11$ ppm.

(In a normal distribution, the interquartile range would be equal to 1.35 times the standard deviation.)

A **95% confidence interval for the median** may be estimated by the observations whose orders in Table 14.2 are:

$$\frac{n + 1}{2} \pm \sqrt{n}$$

where n is the number of observations.

In the present example $n = 25$, so the orders are

$$\frac{25 + 1}{2} \pm \sqrt{25} = 13 \pm 5 = 8 \text{ and } 18$$

The 8th observation in order was 15 ppm, and the 18th was 23 ppm. A 95% confidence interval for the median level of emission is 15 to 23 ppm.

A test for two independent samples: Wilcoxon–Mann–Whitney test

As part of the development of their throat lozenge, Moxon's Medications presented sufferers from sore throats with either Singagain or Hackit, and asked them to report the time it took for one lozenge to take effect. Eight people were given each brand, and the results received from 13 people were as shown in Table 14.3.

A test to compare the effect of the two brands, a non-parametric equivalent of the two-sample t-test, is the Wilcoxon–Mann–Whitney test.

The observations are converted to their ranks, taking all the data together, and the sum of the ranks for each brand obtained. The procedure is shown in Table 14.4.

Table 14.3 Time to take effect

Brand	Times (min.)	Median
Singagain	35, 15, 5, 10, 50, 20, 30	20
Hackit	3, 12, 8, 5, 10, 7	7.5

Table 14.4 Ranking of data

Rank	Time	Brand
1	50	S
2	35	S
3	30	S
4	20	S
5	15	S
6	12	H
7.5	10	H
7.5	10	S
9	8	H
10	7	H
11.5	5	S
11.5	5	H
13	3	H

Notice that the results may be ranked from lowest to highest or vice versa. With this set of data they have been ranked from highest to lowest, so that rank 1 is given to the highest time (50); it was for Singagain; rank 2 was the second highest (35) which was also for Singagain, and so on until the lowest which is given rank 13.

Where there are equal times, these observations share the average of the ranks which would have been given if they had been different; thus ranks 7 and 8 are replaced with rank 7.5.

The sums of ranks for each brand are:

Singagain	$1 + 2 + 3 + 4 + 5 + 7.5 + 11.5$	$= 34$
Hackit	$6 + 7.5 + 9 + 10 + 11.5 + 13$	$= 57$

The significance test is as follows:

Null hypothesis   The median time is the same for both brands.

Alternative hypothesis The median times for the two brands are different.

Test value   The sum of the ranks for the **smaller** sample $= 57$

(There were 6 observations for Hackit, 7 for Singagain.)

Table value   From Table A.14 with sample sizes of 6 and 7:

$56\,1/2$ at the 5% significance level;
$59\,1/2$ at the 1% significance level.

Decision   We can reject the null hypothesis at the 5% level.

Conclusion   Hackit is more effective, having a shorter median time than Singagain.

In this test the order of ranking is important. It was decided for this set of data to rank from highest to lowest values. If we had ranked from lowest to highest the sum of the ranks for Hackit would have been:

$$1 + 2.5 + 4 + 5 + 6.5 + 8 = 27.$$

The test value is the rank sum for the smaller sample, provided ranking is carried out in the direction which makes this as large as possible. We were correct in ranking from highest to lowest times.

A test for paired data: Wilcoxon matched-pairs sign test

In a comparison of the acceptability to dogs of two formulations of Petra dog food, ten dogs were presented with two bowls containing 200 g of either diet. The two

Table 14.5 Amounts of each diet eaten

Diet Dog	A	B	Difference (B−A)
Rover	150	100	−50
Sam	90	80	−10
OJ	120	200	80
Bruno	150	180	30
Sumo	200	200	0
Kevin	125	100	−25
Merton	0	200	200
Pluto	85	100	15
Angel	45	25	−20
Poppet	25	45	20
	Median difference		+7.5

Table 14.6 Ordering the differences

Rank	Difference	Diet eaten more
1	10	A
2	15	B
3.5	20	A
3.5	20	B
5	25	A
6	30	B
7	50	A
8	80	B
9	200	B

formulations were standard (A) and standard with extra jelly (B). After 15 minutes the amount of each that had been eaten was recorded, as shown in Table 14.5.

The nonparametric equivalent of the paired t-test is Wilcoxon's matched-pairs signed ranks test.

The difference (excluding any dogs which indicated no difference) are ranked in order **from lowest to highest** values and the sums of ranks corresponding to each 'direction of difference' are obtained, as in Table 14.6.

The sum of the ranks for each direction was:

A eaten more:	$1 + 3.5 + 5 + 7$	$= 16.5$
B eaten more:	$2 + 3.5 + 6 + 8 + 9$	$= 28.5$

The significance test is as follows:

Null hypothesis	The median difference between amounts eaten of the diets is equal to zero.
Alternative hypothesis	The median difference is not equal to zero.
Test value	The higher of the two rank sums = 28.5.
Table value	From Table A.15
	with sample size = 9 (the number of **non-zero** differences):
	$39\frac{1}{2}$ at the 5% significance level;
	$43\frac{1}{2}$ at the 1% level.
Decision	We cannot reject the null hypothesis.
Conclusion	There is insufficient evidence to conclude that there is a difference between the amounts of each diet eaten.

What type of data can be used?

The tests described in this chapter were demonstrated with data which were measurements which were not assumed to have come from a normal distribution. (If they had, it would have been appropriate and more efficient to have used a two-sample t-test or paired t-test.)

The data were converted to ranks before the test values were calculated.

The data do not need to be measurements. They can originally be in the form of ranks. As long as there is a sense of ordering of the samples, these tests can be used. In fact, these methods are ideal when it is not possible to assign scores or ratings and data are ranked in the first place.

In the two examples below there were no measurements or scores assigned to the samples, they were simply ranked in order of a property.

Example: cracking shoes

Dr Thornton is a product development officer with a well-known footwear manufacturer. He is evaluating a 'synthetic leather' material for making shoes. A wearer trial has been completed with two variants of this material, coded A2 and C4. A batch of black shoes was made using A2 and another batch with C4. Each of a number of volunteers wore a pair during working days for a six-month period. They were given A2 or C4 at random. The final evaluation requires a decision on which of the two variants is least susceptible to cracking when used in footwear. The company does not have a physical test method that will **measure** the extent of cracking in a shoe, so a visual examination takes place. Thornton, having decided to undertake the test himself, places all 12 pairs of shoes on the bench and repeatedly

Table 14.7   Results of the wearer trial

A2	A2	A2	A2	A2	C4	A2	C4	C4	A2	C4	C4
Least cracking											Most cracking

Table 14.8   Comparing two materials using ranking

Material	Rankings											Total
A2	1	2	3	4	5		7			10		32
C4						6		8	9		11	12	46

interchanges adjacent pairs until he is satisfied that the whole line is in rank order. In other words, the most cracked pair of shoes is at one end, the least cracked pair is at the other and there is an increasing degree of cracking along the line.

Throughout this ranking process Thornton has no knowledge of which pairs of shoes were made with each variant of synthetic leather. When he checks the code numbers marked on the underside of the shoes he obtains the ordering shown in Table 14.7.

To reach a decision concerning the two variants of synthetic leather we shall use the Wilcoxon–Mann–Whitney test. We first convert the results in Table 14.7 into rankings. The pair of shoes with least cracking will be given rank 1 and the pair with most cracking given rank 12. The rankings, alongside the material of the pair of shoes, are listed in Table 14.8.

From an inspection of the rankings, Thornton notices that A2 occurs predominantly at the low end of the rankings and C4 is mainly at the top end which would suggest that C4 would be more susceptible to cracking.

Null hypothesis The two variants of synthetic leather are equally susceptible to cracking on average.

Alternative hypothesis The two variants of synthetic leather differ in their susceptibility to cracking on average.

Test value The rank total of the smaller sample $= 46$

Table value From Table A.14 with sample sizes of 5 and 7:
$44\,1/2$ at the 5% significance level;
$48\,1/2$ at the 1% significance level.

Decision Reject the null hypothesis at the 5% significance level.

Conclusion The two variants of synthetic leather differ in their susceptibility to cracking.

Thornton can be confident, therefore, that the A2 material is less susceptible to cracking on average than the C4 material.

Table 14.9   Rankings starting from the most cracking

Material	Rankings											Total
A2	12	11	10	9	8		6		3			59
C4						7		5	4		2	1	19

The direction of the rankings is important in this test. If the ranks had been assigned starting from 1 for the most cracking, the sums would have been quite different, as shown in Table 14.9, which was obtained using the procedure described following Table 14.4.

In the case of Thornton's trial, the rankings in Table 14.8 led to the smaller sample, C4, having a rank total of 46, while the reverse order led to a total of 19. The original ordering was appropriate for the test.

The above experiment resulted in a firm conclusion that shoes made with C4 were more susceptible to cracking than those with A2. Thornton may have been fortunate in that the difference was large enough to detect despite each sample of shoes having a wide variety of wear according to the working and living pattern of each user. Such variability can often mask real differences to be investigated, and it is worthwhile, if possible, to plan an experiment which eliminates such 'nuisance' variation.

After careful consideration of the practical difficulties involved, Thornton decides to adopt a different type of experiment in his next investigation to compare two new materials, L3 and M5. Sixteen volunteers are recruited and a special pair of shoes is made for each one. The pairs are unusual in that one shoe is made from material L3 and the other from M5. In eight of the 16 pairs of shoes the right shoe is made from L3 whilst in the other eight pairs the left shoe is made from this material. Random numbers are used in this allocation so that each volunteer has a 50% chance of having an L3 shoe assigned to his right foot. The participants are not told that the shoes differ. They are simply requested to wear the shoes during working days for a six-month period.

Four of the participants fail to complete the wearer trials, for reasons not connected with the shoes. At the end of the six-month period, therefore, Thornton has 12 pairs of shoes to assess for cracking. With each pair he is primarily concerned with the **difference between the two shoes**. If he finds with several pairs of shoes that there is a difference in cracking, and that the more cracked of the two shoes is made from one particular material in most pairs, then he would have good grounds for concluding that the material is inferior to the other.

As in the previous experiment, Thornton decides to **rank** the pairs of shoes to indicate the degree of cracking. He lines up the pairs of shoes on his bench and labels them A, B, . . . , L. Then he interchanges pairs until he is satisfied that the pair with the **smallest difference** is at one end and the pair with the **largest difference** is at the other end, with an increasing difference along the line. When this rearrangement is complete the order of the pairs of shoes is that shown in Table 14.10.

Table 14.10   Results of second wearer trial

C	H	E	J	A	L	F	G	K	B	I	D

| Smallest difference between shoes in the same pair | | | | | | Largest difference between shoes in the same pair | | | | | |

Table 14.11   Comparing two materials using ranked differences

Material of more cracked shoe				Rankings								Total
L3			3			6	7	8	9	10	11	12	66
M5	1	2		4	5								12

Whilst carrying out this ranking Thornton has no idea which material each shoe is made of. By reference to his records, however, he can easily discover which of the two materials was assigned to the shoe that was more cracked in each pair. He consults his notes and writes on the bench next to each pair of shoes the name of the material that was **more** cracked. He gives a ranking to each pair of shoes, starting with rank 1 for the pair with the smallest difference. The rankings are given in Table 14.11.

Comparing Tables 14.10 and 14.11 we see that, in the pair of shoes with the smallest difference in cracking, pair C, the shoe made from M5 material was more cracked than the shoe made from L3. At the other extreme, in pair D there was the largest difference between the two shoes and it was the one made from L3 that was more cracked.

There is a strong suggestion in Table 14.11 that the L3 material is inferior to M5. We shall test this using the Wilcoxon matched-pairs signed ranks test:

Null hypothesis   The two materials are equally susceptible to cracking on average.

Alternative hypothesis   The two materials differ in their susceptibility to cracking on average.

Test value   The larger of the two rank totals = 66.

Table value   From Table A.15, with a sample size of 12:

$64\frac{1}{2}$ at the 5% significance level;
$70\frac{1}{2}$ at the 1% significance level.

Decision   Reject the null hypothesis at the 5% significance level.

Conclusion   The two materials do differ in their susceptibility to cracking.

This test is similar to the Wilcoxon–Mann–Whitney test carried out earlier except that the data on this occasion consisted of ranked **differences**. Each difference came from a pair of items (shoes) that were **matched** except for the material from which they were made. When we use the test to compare the two materials we know that **within each pair** the 'nuisance variable', extent and roughness of usage, has been equalised. The matched-pairs test is a non-parametric test in that no assumptions about the population distribution are built into the statistical table.

Problems

14.1 To compare the penetration of two wood preservatives, 16 planks of wood were treated with one of the formulations, the planks being allocated at random to each formulation. The planks were visually assessed and the planks placed in order of degree of penetration (1 - penetrated deepest). The formulation used for each plank was as follows:

Order	1	2	3	4	5	6	7	8	9	10	11	12	13	14	15	16
Formulation used	A	A	A	A	B	B	A	A	B	A	B	A	B	B	B	B

Determine by using a significance test whether there is a difference between the formulations in their degree of penetration.

14.2 Fledgeling Personal Products conducted some trials to compare the effectiveness of their anti-dandruff shampoo (F) with that of the established market leader (E). They used the half-head test on 12 dandruff sufferers from among their employees. Each subject had half his or her head washed with one of the shampoos and the other half washed with the other. The shampoos were randomly allocated to the left or right half. The head research officer then assigned a score, up to 20, to indicate the difference between the amounts of dandruff remaining in the two halves of each subject's hair. Larger scores indicate greater differences. She also noted which shampoo had been more effective. The results were as follows:

Subject	Score	More effective shampoo
1	8	F
2	3	F
3	15	E
4	5	E
5	18	E
6	7	F
7	10	E
8	11	F
9	11	E
10	16	E
11	9	E
12	12	E

Use Wilcoxon's matched-pairs test to determine which, if either, of the shampoos is more effective in treating dandruff.

15

Analysis of variance: Components of variability

Introduction

All processes are subject to error. All measurements are also subject to error. One of the initial tasks in an investigation is to obtain estimates of the relative magnitude of these errors so that appropriate action can be taken. For example, there is little point in improving the process if the majority of the variability is due to the test method. Of course, when we obtain a set of test results from a product it is not easy to ascertain how much of the variability in the product is due to the process and how much is due to the test method. However, there is a statistical technique – analysis of variance (ANOVA) – which can achieve this separation.

Overall variability

Figures 15.1–15.4 take us through different stages in obtaining the overall variability from a batch of crisps in which a sample of 10 packets is taken and 10 determinations made on each packet.

Figure 15.1 shows no variability at all either in the test method or in the process – an ideal situation (which may never occur!).

Figure 15.2 shows a situation where there is no test variability but there is process variability. We see each of the packets has a different mean but all ten results from the same packet are the same. The blob diagram is somewhat disjointed since there are only ten possible values but each value is replicated ten times.

In Figure 15.3 we see a situation where there is no process variability but much test variability. In this situation all 100 values are different and they form a blob

Statistical Methods in Practice: for Scientists and Technologists R. Boddy, G. Smith,
© 2009 John Wiley & Sons, Ltd

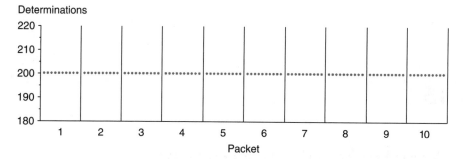

Figure 15.1   Testing SD = 0, Packet SD = 0

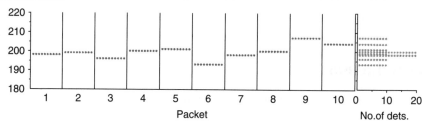

Figure 15.2   Testing SD = 0, Packet SD = 4

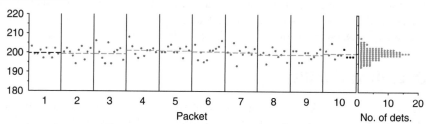

Figure 15.3   Testing SD = 3, Packet SD = 0

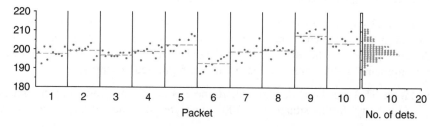

Figure 15.4   Testing SD = 3, Packet SD = 4

Table 15.1   Analysis of variance of data in Figure 15.4

Source of variability	Sum of squares	Degrees of freedom	Mean square	SD
Due to packets	1521	9	169	4
Due to testing	810	90	9	3
Total	2331	99		5

diagram which is close to a normal distribution. We also notice that **although there is no process variability the packet means are all different**. This clearly must be due to the test error and is a complicating factor.

Finally, let us look at Figure 15.4. In this there is both process and testing variability. This is the realistic situation. We see that packet means are much more widely spread than in Figures 15.1–15.3, and the blob diagram also has a much greater range.

Analysis of variance

Table 15.1 shows the analysis of variance of the data in Figure 15.4. We shall explain the terminology and method of calculation later, merely drawing attention, for the time being, to the main feature of analysis: we have managed to separate the error into its two components – 'due to packets' and 'due to testing'.

A practical example

Smallfry is a potato-based snack produced by rolling out potato paste, cutting into minute chips and frying. The company is dissatisfied about the capability of the Smallfry process, that is, the variability is so large that it cannot meet a specification.

To control the quality of Smallfry it is necessary to control the oil. Before a control scheme is instigated it is decided to investigate the amount of variability due to testing compared with the packet-to-packet variability by selecting 10 packets, at random, from a carton of Smallfry, crushing the contents of each packet and then selecting four samples from within each packet. By crushing the contents, the four samples from each packet should be identical and therefore the variation can be ascribed to test error. The results of the investigation are given in Table 15.2.

As we inspect the data, we notice first of all that there are only three determinations from packets 2 and 8. Why was that? Were only three determinations made? Were samples lost? Or were the determinations outliers and they were excluded before being presented for analysis? It is important that all the data be made available so that the statistical analyst can determine the quality of the data.

Table 15.2   Percentage oil for samples from ten packets

Packet no.	1	2	3	4	5	6	7	8	9	10
% Oil	26.7	26.6	25.8	27.4	27.8	27.8	27.9	27.5	28.0	27.1
	26.4	27.2	26.6	26.7	26.8	26.7	27.9	28.3	27.4	27.7
	27.2	27.2	26.3	27.0	26.7	28.1	26.9	26.7	27.2	26.6
	25.7		26.5	26.5	27.1	26.6	26.9		27.8	25.4
Mean	26.5	27.0	26.3	26.9	27.1	27.3	27.4	27.5	27.6	26.7
SD	0.63	0.35	0.36	0.39	0.50	0.76	0.58	0.80	0.37	0.98

Overall mean = 27.02; Overall SD = 0.6750

Table 15.3   Analysis of variance for percentage oil data

Source of variability	Sum of squares	Degrees of freedom	Mean square	SD
Due to packets	6.58	9	0.73	0.31
Due to testing	10.28	28	0.37	0.61
Total	16.86	37		

We can see that there is test-to-test variability as the four (or three) results from a packet are not all the same, with standard deviations ranging from 0.35 to 0.98. There is also possible packet-to-packet variation as the packet means range from 26.3 to 27.6. Whether or not this is more than would be expected from the test error remains to be seen.

We can represent these values in an analysis of variance table as shown in Table 15.3. What can be concluded about the Smallfry process?

The only parts of the table that are of interest to the company are the sources of variability and the standard deviations. They see that the variability is dominated by the test variability, the test-to-test standard deviation being twice as large as the packet-to-packet standard deviation. If they wish to have better control of a process they must have a precise test method. Although they have mitigated this problem on this example by taking four tests per packet, clearly this is impractical in the future. The simple advice is that the test method must first be improved before any improvement can be made to the process.

We shall now see how the ANOVA table is constructed.

Terminology

The **sum of squares** is the sum of squared deviations from an appropriate mean. Thus the total sum of squares is the sum of squared deviations from the overall

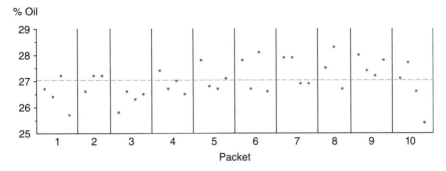

Figure 15.5   Total sum of squares

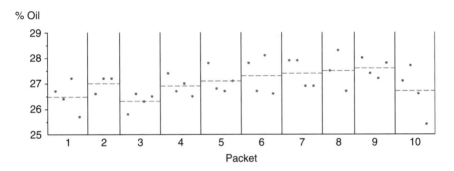

Figure 15.6   Due to testing sum of squares

mean. Its derivation is shown in Figure 15.5 where the squared deviations from the **overall** mean represent the total sum of squares.

The 'due to testing' sum of squares is obtained from the squared deviations between each value and the mean for its packet. This is shown in Figure 15.6.

The total sum of squares is split up by analysis of variance into its two parts, 'due to packets' sum of squares and 'due to testing' sum of squares.

The 'due to packets' sum of squares can therefore be found by subtracting the 'due to testing' sum of squares from the total.

The **mean square** is an intermediate statistic obtained by dividing a sum of squares by its degrees of freedom.

Calculations

Needless to say, there are short-cut methods of calculating sums of squares. These are shown in the calculations below.

First, we must check the degrees of freedom, which follow the same logic as when we calculated standard deviations before:

Total degrees of freedom

$$= \text{total number of observations} - 1 = 38 - 1 = 37$$

'Due to packets' degrees of freedom

$$= \text{number of packets} - 1 = 10 - 1 = 9$$

'Due to testing' degrees of freedom

$$= \sum (\text{number of tests} - 1) = 3 + 2 + 3 + 3 + 3 + 3 + 3 + 2 + 3 + 3$$

$$= 28$$

$$\text{Total sum of squares} = (\text{Total d.f.}) \times (\text{SD of all observations})^2$$

$$= 37 \times 0.6750^2$$

$$= 16.86$$

The 'due to testing' sum of squares is the sum of the squared deviations for each observation about its packet mean.

'Due to testing' sum of squares

$$= \sum [(\text{d.f. for each packet}) \times (\text{SD of determinations in each packet})^2]$$

$$= 3 \times 0.63^2 + 2 \times 0.35^2 + \cdots + 3 \times 0.98^2$$

$$= 10.28$$

The 'due to packets' sum of squares is found by subtracting the 'due to testing' sum of squares from the total, and represents the deviations of the packet means from the overall mean.

'Due to packets' sum of squares

$$= \text{Total sum of squares} - \text{'Due to testing' sum of squares}$$

$$= 16.86 - 10.28$$

$$= 6.58$$

For each component in the ANOVA table, the mean square is determined by dividing the sum of squares by the degrees of freedom:

$$\text{Mean square due to packets} = \frac{\text{'Due to packets' sum of squares}}{\text{'Due to packets' degrees of freedom}}$$

$$= \frac{6.58}{9}$$

$$= 0.73$$

$$\text{Mean square due to testing} = \frac{\text{'Due to testing' sum of squares}}{\text{'Due to testing' degrees of freedom}}$$

$$= \frac{10.28}{28}$$

$$= 0.37$$

To determine the component standard deviations, the test-to-test standard deviation is straightforward, just the square root of the test-to-test mean square.

To obtain the packet-to-packet standard deviation, we need to recognise that the packet-to-packet sum of squares includes a contribution from test error, which we need to subtract before calculating the true packet-to-packet standard deviation. Then, since we are concerned with the variation between packet means, we need to divide by the number of determinations per packet.

Test-to-test SD

$$= \sqrt{\text{'Due to testing' mean square}}$$

$$= \sqrt{0.37}$$

$$= 0.61$$

Packet-to-packet SD

$$= \sqrt{\frac{\text{'Due to packets' mean square} - \text{'Due to testing' mean square}}{\text{Average number of tests per packet}}}$$

$$= \sqrt{\frac{0.73 - 0.37}{3.8}}$$

$$= 0.31$$

Crunch output

Significance Level	5.0%

One - Way Analysis of Variance Tabel							
Source	Sum of Squares	df	Mean Square	SD	F	Table Value	P
Between Packets	6.58	9	0.73	0.31	1.99	2.24	0.079
Within Packets	10.28	28	0.37	0.61			
Total	16.86	37		0.68			

Adjust Rounding	⏶⏷		⏶⏷	⏶⏷

Adjust Rounding	⏶⏷									
Mean	26.5	27	26.3	26.9	27.1	27.3	27.4	27.5	27.6	26.7
Standard Deviation	0.63	0.35	0.36	0.39	0.5	0.76	0.58	0.8	0.37	0.98
Observations	4	3	4	4	4	4	4	3	4	4

	Data Entry (upto 40 batches) >>>									
	Packet									
	1	2	3	4	5	6	7	8	9	10
Determination	26.7	26.6	25.8	27.4	27.8	27.8	27.9	27.5	28	27.1
	26.4	27.2	26.6	26.7	26.8	26.7	27.9	28.3	27.4	27.7
	27.2	27.2	26.3	27	26.7	28.1	26.9	26.7	27.2	26.6
	25.7		26.5	26.5	27.1	26.6	26.9		27.8	25.4

Significance test

The output from *Crunch* includes some columns which we do not require in this example. They relate to a test that the packet-to-packet variation is statistically significant. The test is described and used in the next example in this chapter.

Variation less than chance?

When we determined the packet-to-packet standard deviation, we subtracted the 'due to testing' mean square from the 'due to packets' mean square before dividing by the number of tests per packet and taking the square root.

This was because it is assumed that when there is true packet-to-packet variation it will cause the 'due to packets' mean square to be greater than the 'due to testing' mean square, whereas if there were no packet-to-packet variation the two mean squares would be equal and the packet-to-packet standard deviation would be zero.

It can sometimes happen when there is no packet-to-packet variation that with a set of data the 'due to packets' mean square comes out to be **less than** the 'due to testing' mean square, and the packet-to-packet standard deviation cannot be calculated as it would involve the square root of a negative number.

Crunch displays an error code. The packet-to-packet standard deviation should then be taken as zero.

In such a situation, the data should be examined carefully. Were the packets not selected at random, so that their means were closer together than would be expected from the magnitude of the test error? Was the test error unusually large? Was there an outlier present?

When should the above methods *not* be used?

When there is an outlier in the data. Clearly this should be removed, otherwise it will distort the analysis.

When the test errors are very skewed and there are only a few tests per packet.

Between- and within-batch variability

Worldwide Fish Distributors (UK) have received their first consignment of frozen prawns and their quality control manager, Sid Hughes, wishes to determine the bacteriological quality of the consignment. The consignment is in the form of several batches, and it is possible that there will be variation between batches and between prawns within a batch. He selects four batches at random, thaws them and performs bacteriological analysis on samples of the meat of five prawns from each batch.

It is common practice to use the logarithms of the total viable count (TVC).

The results of the determinations (log TVC, cfu/g) are shown in Table 15.4.

Now Mr Hughes's first impression is that there are large differences between batches, but a colleague urges caution since the sample-to-sample variability within a batch is also high and it is perhaps this which is causing the batch means to vary. A way to overcome this difficulty is to use ANOVA in order to sort out the contribution of each source of variability.

The contributions are:

(i) within-batch variability (between samples only);
(ii) batch-to-batch variability.

The ANOVA table is given in Table 15.5.

A significance test of batch-to-batch variability is carried out as follows:

Null hypothesis   There is no batch-to-batch variability over and above within-batch variability.

Table 15.4   Results of bacteriological tests (log TVC)

Batch	Sample					Mean	SD
	1	2	3	4	5		
A	3.21	2.93	3.09	2.53	2.94	2.94	0.2567
B	4.33	4.55	4.18	3.53	5.06	4.33	0.5576
C	3.93	3.77	3.11	2.85	3.24	3.38	0.4550
D	2.93	4.23	3.56	4.08	3.15	3.59	0.5656

Mean of all data = 3.56; SD of all data = 0.6754; SD of batch means = 0.5804

Table 15.5   Analysis of variance table

Source of variation	Sum of squares	Degrees of freedom	Mean square	Test value	SD
Between batches	5.052	3	1.684	7.45	0.54
Within batch	3.615	16	0.226		0.48
Total	8.667	19			

Alternative hypothesis   There is batch-to-batch variability over and above within-batch variability.

Test value

$$= \frac{\text{Between-batches mean square}}{\text{Within-batch mean square}}$$

$$= \frac{1.684}{0.226}$$

$$= 7.45$$

Table value   From Table A.6 (**one-sided**) with 3 and 16 degrees of freedom:
3.24 at the 5% significance level;
5.29 at the 1% significance level.

Decision   Since the test value is greater than the table value at the 1% significance level we can reject the null hypothesis.

Clearly batch-to-batch variability does exist and we can now estimate the standard deviations attributable to the two sources. These are:

The **within-batch standard deviation** (residual standard deviation) is due to random errors which usually arise from a number of causes. In this situation errors could be due to biological differences between prawns, dispersion of bacteria, errors in the counting method.

The **batch-to-batch standard deviation** is the extra variability due to systematic errors from batch to batch probably caused, in this case, by differences in quality of processing from batch to batch.

Within-batch SD (s_w)

$$= \sqrt{\text{Within-batch mean square}}$$

$$= \sqrt{0.226}$$

$$= 0.48 \text{ log number}$$

Batch-to-batch SD (s_b)

$$= \sqrt{\frac{\text{Between-batches mean square} - \text{Within-batch mean square}}{\text{Number of samples per batch}}}$$

$$= \sqrt{\frac{1.684 - 0.226}{5}}$$

$$= 0.54 \text{ log number}$$

Hughes has obtained estimates of both standard deviations. They are both similar, being about half a log number, but it must be emphasised that the batch-to-batch SD will be an imprecise estimate, being based on only four batches, while the within-batch SD is a far better estimate, based on 20 observations. His analysis is appropriate as long as:

(a) the prawns are a random selection from each selected batch;
(b) the four batches are representative of the population of batches in the consignment.

How many batches and how many prawns should be sampled?

With this knowledge of the two standard deviations representing how the variability occurs in the consignment, it is possible to determine which sampling regime will give a suitable precision of the mean bacteriological quality of the batch.

The **precision** (or standard error) of the mean log TVC is given by:

$$\sqrt{\frac{(\text{Batch-to-batch SD})^2}{\text{No. of batches}} + \frac{(\text{Within-batch SD})^2}{(\text{No. of batches}) \times (\text{No. of samples per batch})}}$$

In the above example, four batches were sampled, and five prawns per batch, so the precision of the mean is:

$$\sqrt{\frac{0.54^2}{4} + \frac{0.48^2}{4 \times 5}} = 0.29 \text{ log number}$$

If it is desired to reduce the precision of the mean, more data will be needed.

But should we take more samples per batch or more batches?

Hughes tries out some sampling regimes and obtains the values for the precision shown in Table 15.6.

It is clear that increasing the number of samples per batch will make little impression on the width of the confidence interval, and the answer lies in increasing the

Table 15.6   Precision of the mean with different sampling regimes

No. of batches	Number of samples per batch	Total number of observations	Precision of mean
4	5	20	0.29
4	10	40	0.28
8	5	40	0.21
8	3	24	0.21
8	2	16	0.23
16	1	16	0.18

number of batches. At the same time, work can be saved by reducing the number of samples per batch.

This conclusion applies in this case because the variation between batches was greater than the variation within a batch. Different conclusions would be reached if the major variation occurred between samples within a batch, and it is wise to carry out some exploratory calculations to see the effect of different sampling regimes on the width of the confidence interval.

Crunch output

Significance Level	5.0%

One - Way Analysis of Variance Table

Source	Sum of Squares	df	Mean Square	SD	F	Table Value	P
Between Batches	5.05	3	1.68	0.54	7.45	3.24	0.002
Within Batches	3.62	16	0.23	0.48			
Total	8.67	19		0.72			

Adjust Rounding	▲▼		▲▼	▲▼			

Adjust Rounding	▲▼						
Mean	2.94	4.33	3.38	3.59			
Standard Deviation	0.257	0.558	0.455	0.566			
Observations	5	5	5	5			

Data Entry (up to 40 batches) >>>

Batch							
A	B	C	D	5	6	7
Prawn	3.21	4.33	3.93	2.93			
	2.93	4.55	3.77	4.23			
	3.09	4.18	3.11	3.56			
	2.53	3.53	2.85	4.08			
	2.94	5.06	3.24	3.15			

Precision of Mean Calculator

No. of Batches	4
No. of Prawns per Batch	5
SD (Precision of Mean)	0.29

Problems

15.1 The Motor Industry Revolutionary Research Organisation (MIRROR) run testing services for their clients in which cars are placed on a rolling road (consisting of two sets of idling rollers – one for the front wheels and one for the back wheels) and driven under various conditions to resemble road running. The advantage of the rolling road is that the performance can be closely controlled and monitored for numerous performance measures. One of their frequent tests is to monitor the fuel economy of different oils using the same car and driving conditions.

Recently their results have become far too variable, and to narrow the search for the cause of the variability they need to know whether it is due to:

1. daily setting-up conditions including calibration of the rolling road;
2. setting-up conditions for each test including temperature stabilisation of the engine.

Each day they carry out nine tests, six on candidate oils and three reference tests on the same oil with reference tests interspersed among the candidate oils. The last eight days of testing yielded the following results for fuel economy (miles per gallon) on the reference tests:

Day 1	Day 2	Day 3	Day 4	Day 5	Day 6	Day 7	Day 8
32.31	35.39	31.52	34.63	31.19	33.32	33.24	34.50
33.79	33.24	33.05	36.67	32.33	33.57	35.50	35.13
32.13	34.11	31.46	35.98	31.56	34.50	34.75	35.60

(a) Fill in the blank cells in the table below:

Source of Variability	Sum of Squares	Degrees of Freedom	Mean Square
	To be calculated		To be calculated
	To be calculated		To be calculated
	To be calculated		

Using *Crunch*, enter the data into the 'ANOVA' sheet.
Answer the following questions in order to familiarise yourself with the output:

(b) Which day gave the largest fuel economy and had the highest mean?
(c) What is the standard deviation for day 4?
Check that the output agrees with your answer to the degrees of freedom in part (a). If it does not, it is likely that you have requested the wrong

address for the data or put the data in the wrong way round. (Always check the degrees of freedom with your hand calculations.)

(d) Notice the p-value. If this is less than 0.05 it can be inferred that between-days variability is significant at the 5% level. What is the p-value and what conclusion can be drawn?

(e) What are the values of the within-day standard deviation and the day-to-day standard deviation?

(f) Calculate the overall standard deviation which equals:

$$\sqrt{SD^2_{\text{between days}} + SD^2_{\text{within day}}}$$

(g) Which part of the test procedure should be investigated further?

15.2 Cornabisc buy large quantities of corn to produce their 'top-of-the-range' breakfast cereal. The quality of corn arriving at the factory causes problems in blending to obtain a satisfactory raw material for feeding into the process. They have, therefore, a system in which payment for each lot depends upon the quantity of protein. However, it is difficult to obtain precise estimates since the delivery is highly variable and also the test method is far from accurate. They have, however, carried out an experiment by obtaining several 25 g samples from a delivery and subjecting each one to multiple determinations. The results are as follows:

Sample	A	B	C	D	E
	120	88	104	106	126
	112	94	98	118	119
	113		86	107	139
				121	118
					133

(a) Fill in the blank cells in the table below:

Source of Variability	Sum of Squares	Degrees of Freedom	Mean Square
	To be calculated		To be calculated
	To be calculated		To be calculated
	To be calculated		

Using *Crunch*, enter the data into the 'ANOVA' sheet.
Check that the output agrees with your answer for degrees of freedom in part (a).

Answer the following questions:

(b) What are the values of the test-to-test SD and the sample-to-sample SD?

(c) Which component of variation is more likely to influence the overall variability?

(d) Based on the above results, Cornabisc wish to devise a strategy which gives them the best precision for the mean value. This will depend upon the number of samples and the number of tests per sample.

SD (for the precision of the mean) =

$$\sqrt{\frac{(\text{Sample-to-sample SD})^2}{\text{No. of samples}} + \frac{(\text{Test-to-test SD})^2}{(\text{No. of samples}) \times (\text{No. of tests per sample})}}$$

It is five times as costly to take a sample as to carry out a test. The strategies below are approximately equally expensive. Calculate the standard deviation (for the precision of the mean) for each strategy and decide which one is preferable:

(i) 2 samples, 13 tests per sample;

(ii) 5 samples, 2 tests per sample;

(iii) 6 samples, 1 test per sample.

16

Cusum analysis for detecting process changes

Introduction

Vast amounts of data are collected on a regular basis, such as: production records on yield, impurity, specification parameters; calibration and quality control samples from a test method; numbers of customer complaints per week; response times for a computer system; demand and sales records. Any data collected on a regular basis is suitable for cumulative sum analysis, which helps us to decide:

(a) if changes have taken place;
(b) approximately when changes took place.

The technique will be illustrated using an example from a lamp producer.

Analysing past data

Globrightly produce studio lamps. Currently, they are concerned about two issues – cost and quality – relating to their CY lamp. Ed Eddie, the owner of the company, has told his son to investigate. His initial investigation highlights two problems concerned with yield and light intensity.

Yield is defined as the percentage of glass bulbs entering the process which are successfully built into completed lamps. This is calculated on a daily basis. Clearly low yield represents a large loss to the company. For CY lamps, the light intensity is of prime concern since any change in intensity causes problems for their major customers on film sets and in photographic studios. Eddie's son finds that each day a sample of five lamps is taken from production, their light intensity measured and the mean recorded.

Initially he decides to concentrate on mean intensity for each day.

Statistical Methods in Practice: for Scientists and Technologists   R. Boddy, G. Smith,
© 2009 John Wiley & Sons, Ltd

Intensity

Eddie's son decides his next step is to analyse past data from the previous 50 days of manufacture. This data set is given in Table 16.1. For the time being, we will concentrate on the intensity data.

In Figure 16.1 we see the raw intensity data plotted from day to day, usually referred to as a runs chart. There is clearly a considerable amount of variation from day to day and it is difficult to detect any trends.

Figure 16.2 shows an alternative presentation of the same data. When we examine it we are struck most forcibly by a very clear pattern. This same pattern is actually embedded in Figure 16.1, but it does not stand out so well in the standard plot. Though Figures 16.1 and 16.2 transmit the same message, it is more easily perceived from the latter.

Table 16.1   Results from the previous 50 days of production

Day	1	2	3	4	5	6	7	8	9	10	11	12	13
Intensity	253	246	259	254	248	261	251	240	256	249	253	257	251
Yield	95.9	92.8	93.7	96.7	93.2	93.6	95.0	97.1	95.4	94.6	93.0	95.1	97.0

14	15	16	17	18	19	20	21	22	23	24	25	26
253	244	260	250	255	245	248	259	250	241	255	252	263
98.1	96.8	95.0	97.9	95.8	97.8	98.5	95.9	97.5	98.3	96.2	93.7	96.3

27	28	29	30	31	32	33	34	35	36	37	38	39
252	248	261	257	250	255	240	248	244	247	252	243	239
95.2	95.1	97.2	98.4	98.3	98.3	95.7	96.1	94.5	96.7	94.8	96.4	95.6

40	41	42	43	44	45	46	47	48	49	50	Mean
242	256	238	244	249	258	243	248	246	239	248	250
94.9	97.9	95.2	93.1	94.8	96.8	95.3	96.2	97.0	96.0	94.6	95.9

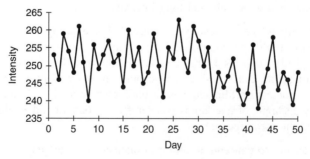

Figure 16.1   Runs plot of light intensity

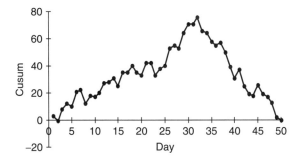

Figure 16.2   Cusum plot of light intensity

Two questions need to be answered at this point:

(a) What have we done to the data to convert Figure 16.1 into Figure 16.2?
(b) How are we to interpret Figure 16.2?

The calculations carried out in order to transform the raw data of Figure 16.1 into the **cusums** of Figure 16.2 are set out in Table 16.2. Two steps are involved: first we subtract a **reference value** from each observation to obtain the deviations in column 3; then we add up these 'deviations from reference value' in order to obtain the cusums in column 4. The cusum column is simply the **cumulative sum** of the deviation column. Figure 16.2 is a graph of column 4, plotted against the day in column 1. For historical data it is convenient to use the mean of all the data in the analysis as the reference value, but if there is an actual target value it would be appropriate to use it.

The reference value is 250, the mean of the 50 observations. A consequence of using the mean as the reference value is that the final entry in the cusum column is zero, and hence the cusum graph **starts and finishes on the horizontal axis**. It is possible that a cusum plot will wander up and down in a random manner, never deviating far from the horizontal axis. It may, on the other hand, move well away from the axis only to return later. If this is the case there will be marked **changes in the slope** of this plot. Visual inspection of Figure 16.2 might lead us to conclude the slope changes, at about day 32. This **change point** appears to split the series in two:

(a) from day 1 to day 32 the graph has a **positive slope**;
(b) from day 33 to day 50 the graph has a **negative slope**.

The slope of a stage of the cusum graph is related to the mean intensity for the days in that stage. Thus the **positive** slope for days 1 to 32 tells us that the mean is **above** the reference value of 250, and the **negative** slope for days 33 to 50 tells us that these days have a mean intensity **below** the **reference** value.

We shall return to this point later. First we must ensure, as far as is reasonably possible, that the change point is real and not just due to simple random patterns,

Table 16.2   Calculations of cusums from intensity data

Day	Intensity	Deviation from reference Value	Cusum	Day	Intensity	Deviation from reference Value	Cusum
1	253	3	3	26	263	13	53
2	246	−4	−1	27	252	2	55
3	259	9	8	28	248	−2	53
4	254	4	12	29	261	11	64
5	248	−2	10	30	257	7	71
6	261	11	21	31	250	0	71
7	251	1	22	32	255	5	76
8	240	−10	12	33	240	−10	66
9	256	6	18	34	248	−2	64
10	249	−1	17	35	244	−6	58
11	253	3	20	36	247	−3	55
12	257	7	27	37	252	2	57
13	251	1	28	38	243	−7	50
14	253	3	31	39	239	−11	39
15	244	−6	25	40	242	−8	31
16	260	10	35	41	256	6	37
17	250	0	35	42	238	−12	25
18	255	5	40	43	244	−6	19
19	245	−5	35	44	249	−1	18
20	248	−2	33	45	258	8	26
21	259	9	42	46	243	−7	19
22	250	0	42	47	248	−2	17
23	241	−9	33	48	246	−4	13
24	255	5	38	49	239	−11	2
25	252	2	40	50	248	−2	0

Reference value = 250

for, in any series, we shall see points at which the slope appears to change. Our task is now to see whether the change point is significant.

To carry out the significance test we must have a measure of variability as represented by the error from day to day. What should we use?

(i) The standard deviation of the 50 days' intensities? No, this would not truly represent day-to-day error since it would be inflated by the presence of step changes in the data.

(ii) A combined standard deviation obtained by combining the standard deviations from each stage between step changes? This would be ideal, but we cannot do this until we have established when the true step changes occurred.

(iii) We overcome both problems by using the **localised standard deviation**, which is equivalent to combining the standard deviations of each pair of consecutive days' observations, thus minimising the effect of step changes.

The value of the localised standard deviation for this data set is 7.22. The calculation of the localised standard deviation is shown below.

Localised standard deviation

The first stage is the calculation of the successive differences – that is, the absolute difference between each observation and the previous one. These are shown in Table 16.3.

Table 16.3   Computation of localised standard deviation

Day	Intensity	Difference	Day	Intensity	Difference
1	253		26	263	11
2	246	7	27	252	11
3	259	13	28	248	4
4	254	5	29	261	13
5	248	6	30	257	4
6	261	13	31	250	7
7	251	10	32	255	5
8	240	11	33	240	15
9	256	16	34	248	8
10	249	7	35	244	4
11	253	4	36	247	3
12	257	4	37	252	5
13	251	6	38	243	9
14	253	2	39	239	4
15	244	9	40	242	3
16	260	16	41	256	14
17	250	10	42	238	18
18	255	5	43	244	6
19	245	10	44	249	5
20	248	3	45	258	9
21	259	11	46	243	15
22	250	9	47	248	5
23	241	9	48	246	2
24	255	14	49	239	7
25	252	3	50	248	9
			Mean		8.143

To obtain the localised standard deviation the mean absolute difference is divided by 1.128

$$\text{Localised SD} = \frac{\text{Mean difference}}{1.128}$$

$$= \frac{8.143}{1.128}$$

$$= 7.22$$

Significance test

We are now in a position to carry out a significance test, which is carried out as follows:

Null hypothesis   There was no change in the process mean intensity during the 50 days.

Alternative hypothesis   There was a change in the process mean intensity during this period.

Test value $= \dfrac{\text{Maximum }|\text{cusum}|}{\text{Localised SD}}$

$$= \frac{76}{7.22}$$

$$= 10.5$$

Table values   For a span of 50 observations, Table A.9 gives

8.6 at the 5% level of significance;
10.4 at the 1% level of significance.

Decision   As the test value is greater than the 1% table value we reject the null hypothesis.

Conclusion   We conclude that a change of mean intensity **did** occur during the period, the maximum cusum being at day 32.

Let us summarise our findings regarding light intensity. We have shown the series can be split into two stages, namely:

Stage	Mean
Days 1 to 32	252.4
Days 33 to 50	245.8

Crunch output

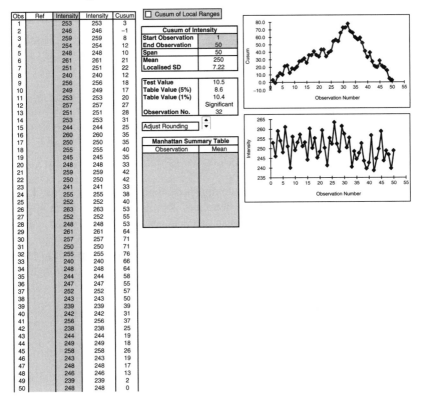

Obs	Ref	Intensity	Intensity	Cusum
1		253	253	3
2		246	246	-1
3		259	259	8
4		254	254	12
5		248	248	10
6		261	261	21
7		251	251	22
8		240	240	12
9		256	256	18
10		249	249	17
11		253	253	20
12		257	257	27
13		251	251	28
14		253	253	31
15		244	244	25
16		260	260	35
17		250	250	35
18		255	255	40
19		245	245	35
20		248	248	33
21		259	259	42
22		250	250	42
23		241	241	33
24		255	255	38
25		252	252	40
26		263	263	53
27		252	252	55
28		248	248	53
29		261	261	64
30		257	257	71
31		250	250	71
32		255	255	76
33		240	240	66
34		248	248	64
35		244	244	58
36		247	247	55
37		252	252	57
38		243	243	50
39		239	239	39
40		242	242	31
41		256	256	37
42		238	238	25
43		244	244	19
44		249	249	18
45		258	258	26
46		243	243	19
47		248	248	17
48		246	246	13
49		239	239	2
50		248	248	0

☐ Cusum of Local Ranges

Cusum of Intensity

Start Observation	1
End Observation	50
Span	50
Mean	250
Localised SD	7.22

Test Value	10.5
Table Value (5%)	8.6
Table Value (1%)	10.4
	Significant
Observation No.	32

Adjust Rounding ▲ ▼

Manhattan Summary Table

Observation	Mean

Yield

Having completed the analysis of intensity, let us look at yield. A runs plot of the data is given in Figure 16.3 and a cusum plot in Figure 16.4.

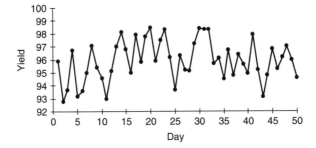

Figure 16.3   Runs plot of yield

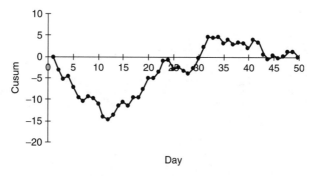

Day

Figure 16.4   Cusum plot of yield

The cusum plot of yield is far more complicated than that observed for light intensity. The significance test applies when there is only one change point within the data set. We therefore need to establish a procedure for analysing data which may contain more than one change point, as in practice there may be many. We shall assume that we have *Crunch* available with a cusum analysis program. The steps in the analysis are given below. A 5% significance level is used throughout the analysis.

Step 1. By visual observation decide on the possible change points.
These are at Days 12, 24, 28 and 32.

Step 2. Set Start Observation 1
End Observation 12.
This will check that there are no significant changes before the first 'observed' change point.
As shown in the *Crunch* output below, mean = 94.7; localised SD = 1.66; test value = 1.3; table value = 4.3 from Table A.9 for a span of 12.
We conclude that there is no significant change point.

Obs	Ref	Yield	Yield	Cusum
1		95.9	95.9	1.2
2		92.8	92.8	−0.6
3		93.7	93.7	−1.6
4		96.7	96.7	0.4
5		93.2	93.2	−1.1
6		93.6	93.6	−2.1
7		95	95	−1.8
8		97.1	97.1	0.6
9		95.4	95.4	1.3
10		94.6	94.6	1.3
11		93	93	−0.4
12		95.1	95.1	0
13		97		
14		98.1		
15		96.8		
16		95		
17		97.9		
18		95.8		
19		97.8		
20		98.5		
21		95.9		
22		97.5		
23		98.3		
24		96.2		
25		93.7		
26		96.3		

☐ Cusum of Local Ranges

Cusum of Yield	
Start Observation	1
End Observation	12
Span	12
Mean	94.7
Localised SD	1.66

Test Value	1.3
Table Value (5%)	4.3
Table Value (1%)	5.3
	Not Sig.
Observation No.	6

Adjust Rounding

Manhattan Summary Table	
Observation	Mean

As we are looking only at the first 12 observations, the start observation is 1, the end observation is 12 and the span (the number of observations in the range of interest) is 12.
The localised standard deviation is recalculated for this span.

Step 3. Because there was no significant change, increase the end observation 12 to the next observed change point. The span is now days 1 to 24. Mean = 95.9; localised SD = 1.60; test value = 9.0; Table value = 6.0.
There is a significant change point. The maximum cusum occurred at day 12.

Step 4. Because there is a significant change point increase the start observation to 13 to give a span of days 13 to 24. Mean = 97.1; localised SD = 1.53; test value = 1.2; table value = 4.3.
There is no significant change point.

Step 5. Increase end observation to 28, next observed change point, to give a span of 13 to 28. Mean = 96.6; localised SD = 1.50; test value = 4.2; table value = 5.0.
There is no significant change point.

Step 6. Increase end observation to 32, the next observed change point, to give a span of 13 to 32. Mean = 96.9; localised SD = 1.34; test value = 3.5; table value = 5.6.
There is no significant change point.

Step 7. Increase end observation to 50, the last point in the series, to give a span of 13 to 50. Mean = 96.3; localised SD = 1.38; test value = 8.4; table value = 7.6.
There is a significant change point. The maximum cusum occurred at day 32.

Step 8. Increase start observation to 33, to give a span of 33 to 50. Mean = 95.6; localised SD = 1.37; test value = 1.5; table value = 5.3.
There is no significant change point.

We can summarise our findings:

Stage	Mean
Days 1–12	94.7
Days 13–32	96.9
Days 33–50	95.6

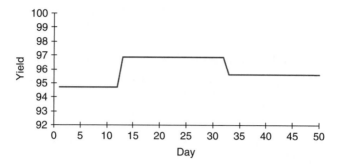

Figure 16.5   Manhattan diagram of yield

We can present this information in the form of a Manhattan diagram as shown in Figure 16.5 (so called because such a graph with many changes will resemble the skyline of a modern city). See also the *Crunch* output below.

Manhattan Summary Table	
Observation	Mean
1	94.7
13	96.9
33	95.6

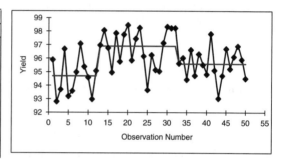

In the table the observation number at the start of each stage and the mean for that stage are entered. Once all stages have been entered, the graph is completed by setting start and end observations for the whole data set.

Conclusions from the analysis

Eddie's son decides to summarise the data as a series of graphs as shown in Figure 16.6.

Eddie's son must now turn his attention to why these step changes occurred. If he can find the reason, it may be possible to eliminate the step changes in intensity, thereby improving quality, and move the yield to the higher level, thereby lowering costs. He decides to search the process logs to examine the occurrences around day 12, which increased yield, and around day 32, which decreased both yield and light intensity. He is immediately rewarded.

(i) The R & D Department had carried out research showing that tensioning of the filament was being set too high. A change in tension on the CY lamps had been introduced at day 11.

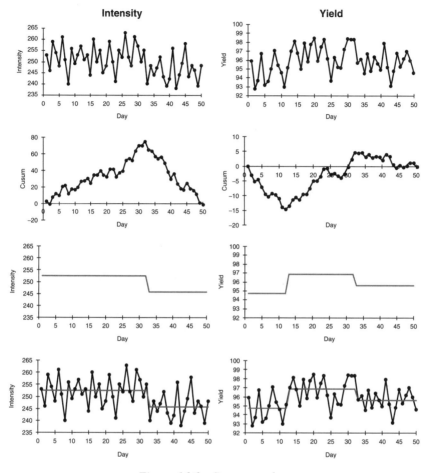

Figure 16.6   Summary plots

(ii) A decision, based on cost, had been made to change to a new supplier of filaments which occurred at day 32. This could be the cause of the changes in light intensity and yield.

Eddie's son decides to keep a careful control of filament tensioning. He also decides to revert back to the old supplier and see if the cusum plot changes to a positive slope.

Problem

16.1 Dr Lightowler has been delegated to find the reason for the variability in the yield of a process. As a first step he decides to examine the previous 60 observations and to concentrate his attention on two parameters, the quantity of catalyst present and the activity of catalyst.

Both these variables are measured by sampling every shift. Varying amounts of catalyst are added at irregular intervals to maintain performance. He also calculates the % yield on a basis which relates to the period of sampling. The results are given below.

Sample no.	1	2	3	4	5	6	7	8	9	10
% Yield	93.3	91.0	95.9	93.2	92.3	95.7	94.0	91.8	92.6	95.1
Catalyst quantity	2420	2420	2430	2380	2380	2440	2420	2400	2430	2410
Catalyst activity	132	139	135	131	139	133	138	132	135	135

11	12	13	14	15	16	17	18	19	20
96.5	91.2	95.0	95.5	95.4	95.9	91.7	90.7	93.3	90.3
2370	2440	2440	2370	2310	2350	2370	2310	2350	2380
139	138	139	133	137	137	136	133	136	134

21	22	23	24	25	26	27	28	29	30
92.3	92.7	93.5	92.7	92.2	92.1	95.8	97.0	94.0	92.2
2330	2350	2310	2330	2360	2420	2380	2440	2460	2460
136	134	132	136	140	141	139	142	138	134

31	32	33	34	35	36	37	38	39	40
97.2	95.7	94.5	93.8	95.8	92.6	94.6	95.1	92.0	92.5
2390	2400	2410	2460	2390	2420	2470	2450	2410	2420
136	141	138	140	141	141	138	137	134	144

41	42	43	44	45	46	47	48	49	50
94.4	96.4	92.2	97.5	96.2	94.1	95.1	93.5	95.3	91.4
2450	2410	2400	2420	2390	2460	2450	2360	2380	2370
138	139	141	141	141	141	139	143	141	139

51	52	53	54	55	56	57	58	59	60
90.5	93.4	89.6	90.8	92.5	93.6	93.9	93.3	91.4	92.0
2350	2390	2330	2380	2410	2330	2330	2390	2350	2310
144	146	145	145	140	146	146	140	146	143

(a) Undertake a cusum analysis of percentage yield. Determine the significant step changes.

(b) Repeat the analysis with catalyst quantity and catalyst activity.

17

Rounding of results

Introduction

When we say a value is rounded to 20, we mean the value is around 20 in some sense. Clearly, rounding data sensibly has many advantages in evaluating data. However, poor rounding has just the opposite effect. Let us look at some examples of poor rounding.

> *"I am working to improve the purity of a drug. Before commencing my work it was 99.8%, now it is 99.9%. (I was always told to round to three significant figures.) Thus impurity has now been halved."*

The actual figures were 99.841% and 99.872%.

> *"Your certificate of analysis, giving repeat determinations of 20, 15 and 20 mg/l is not acceptable."*

In fact the determinations were 17.62, 17.41 and 17.52 rounded to the nearest five.

> *"In January we sold goods to the value of £56,882,412.31, in February £51,047,126.21, in March £49,674,316.23, in April £54,321,307.21. January sales were higher than February's by £5,835,286.10 ..."*

Clearly there are many digits in each figure. It would be easier to understand if the values were quoted as January £56.9 m, February £51.0 m, March £49.7 m and April £54.3 m.

Statistical Methods in Practice: for Scientists and Technologists   R. Boddy, G. Smith,
© 2009 John Wiley & Sons, Ltd

Choosing the rounding scale

Our first decision when rounding is to decide the rounding scale. We have a choice of the evens, fives or tens scale. Table 17.1 illustrates the scales with the same set of five results.

Depending upon the discrimination needed, we can decide, with this scale, whether to round to even numbers only, or odd and even numbers.

By far the most popular scale is the decimal scale and it is this scale which is used in the following example.

Table 17.1   Three different rounding scales

'Tens' Scale			
271.13	271.1	271	270
273.64	273.6	274	270
282.17	282.2	282	280
276.19	276.2	276	280
277.40	277.4	277	280
Rounded to two decimal places	Rounded to one decimal place	Rounded to whole number	Rounded to nearest 10

'Fives' Scale		
271.15	271	270
273.65	273.5	275
282.15	282	280
276.2	276	275
277.4	277.5	275
Rounded to nearest multiple of 0.05	Rounded to nearest multiple of 0.5	Rounded to nearest multiple of 5

'Evens' Scale			
271.1	271.2	271	272
273.6	273.6	274	274
282.2	282.2	282	282
276.2	276.2	276	276
277.4	277.4	277	278
Rounded to nearest multiple of 0.1	Rounded to nearest multiple of 0.2	Rounded to nearest whole number	Rounded to nearest even number

Reporting purposes: deciding the amount of rounding

Ruff Lee is the manager of the QA laboratory for Fibrillated Fibres. One of the main tests his technicians undertake is the tensile strength of fibres. Since the introduction of digital recorders the technicians have, in his opinion, been recording a ridiculous number of figures for each test. Not only are many of the figures meaningless but in Ruff's opinion they may lead to gross errors due to transposition of digits. He requires meaningful rounding so that he can look at a set of figures and make a judgement about them.

The Standard Operating Procedure requires that four samples are tested from each bobbin sampled with 9000 m of fibre unwrapped between each test. Each test is destructive so the replicates include both sampling and testing errors.

In order to decide on a rounding procedure it is often necessary to collect more data than would be obtained from routine analysis. Therefore Ruff instructs one of his technicians to test 10 samples from a single bobbin. He records the following results (N):

26.1704572, 24.5612719, 20.9390421, 23.2213741, 24.1716256,

26.1425314, 29.2170347, 22.4416359, 25.1942413, 24.0134201

and calculates the standard deviation as 2.285368.

Using the decimal scale:

Determine the number of decimal places from the indicator

$$= \frac{SD}{2} \text{(as advised in BS 2846-1)}$$

$$= \frac{2.285368}{2}$$

$$= 1.142684$$

This value for the indicator (referred to as "quotable interval" in BS 2846-1) is above 1 and below 10. We have a choice of rounding to a whole number or the nearest 10. The convention recommended in BS 2846-1 is to choose the lower of the two numbers. The individual results should therefore be reported to the nearest whole number.

Reporting purposes: rounding of means and standard deviations

To round a mean we use the same principle but a slightly different formula. Remember that Ruff will normally test four samples per bobbin. The estimate of the number of decimal places for the mean depends on the number of observations

(n) which are averaged.

$$\text{Indicator} = \frac{\text{SD}}{\sqrt{4n}}$$

$$= \frac{2.283568}{\sqrt{16}}$$

$$= 0.571342$$

The means should therefore be rounded to one decimal place.

To round the standard deviation, again use the same approach but with another formula. The standard deviations should be rounded to one decimal place.

$$\text{Indicator} = \frac{\text{SD}}{\sqrt{4n}}$$

$$= \frac{2.283568}{\sqrt{32}}$$

$$= 0.404000$$

Again the quotable interval is 0.1.

Recording the original data and using means and standard deviations in statistical analysis

The levels of rounding used above are fine for making judgements and reports but lead to inaccuracy when carrying out further statistical analysis. A simple rule is as follows. For recording the original data and for statistical analysis **one more decimal place should be used**.

Thus the original data should be recorded to 0.1 and the mean and standard deviation recorded to 0.01.

The next six batches give the following results. Notice that the tests are recorded to only one decimal place. Table 17.2 shows the data rounded to one decimal place which is suitable for statistical analysis while Table 17.3 shows the data rounded for presentational purposes.

Table 17.2   Data rounded to one decimal place

Bobbin	Data recorded as				For statistical analysis	
					Mean	SD
A	25.3	26.1	23.8	25.9	25.28	1.04
B	25.9	24.2	23.6	22.9	24.15	1.28
C	20.4	19.3	20.6	17.8	19.52	1.28
D	27.9	31.1	25.1	29.2	28.33	2.52
E	16.4	18.6	20.4	15.9	17.83	2.08
F	19.9	25.7	24.2	23.0	23.20	2.46

Table 17.3   Data rounded for presentational purposes

Meaningful rounding						
Bobbin					Mean	SD
A	25	26	24	26	25.3	1.0
B	26	24	24	23	24.2	1.3
C	20	19	21	18	19.5	1.3
D	28	31	25	29	28.3	2.5
E	16	19	20	16	17.8	2.1
F	20	26	24	23	23.2	2.5

Reference

BS 2846-1:1991
Guide to statistical interpretation of data. Routine analysis of quantitative data. British Standards Institution, London.

Solutions to Problems

Chapter 2

2.1 (a) No. of determinations (n) $= 7$
Sample mean $(\bar{x})$ $= 31.66$
Sample standard deviation (s) $= 1.198$ with 6 degrees of freedom

(b) From Table A.2 with 6 degrees of freedom, $t = 2.45$ for the 95% confidence interval. So the 95% confidence interval for the mean is

$$\bar{x} \pm \frac{ts}{\sqrt{n}} = 31.66 \pm \frac{2.45 \times 1.198}{\sqrt{7}}$$

$$= 31.66 \pm 1.11$$

$$= 30.55 \text{ to } 32.77 \, \text{g/l}$$

(c) As part of the confidence interval is below 32 g/l, there is no evidence that the requirement is being met.

(d)
$$n = \left(\frac{ts}{c}\right)^2$$

$$= \left(\frac{2.45 \times 1.198}{0.5}\right)^2$$

$$= 34$$

2.2 (a) Moisture at high speed/high temperature:

$n = 8$, $\bar{x} = 12.3$, $s = 1.070$ with 7 degrees of freedom.

From Table A.2 with 7 degrees of freedom, $t = 2.36$. So the 95% confidence interval for the population mean is

$$\bar{x} \pm \frac{ts}{\sqrt{n}} = 12.3 \pm \frac{2.36 \times 1.070}{\sqrt{8}}$$

$$= 12.3 \pm 0.89$$

Statistical Methods in Practice: for Scientists and Technologists   R. Boddy, G. Smith,
© 2009 John Wiley & Sons, Ltd

The 95% confidence interval for the mean moisture level at high speed and high temperature is 11.4% to 13.2%.
Moisture at normal speed/normal temperature:

$n = 10$, $\bar{x} = 12.1$, $s = 0.965$ with 9 degrees of freedom.

From Table A.2 with 9 degrees of freedom, $t = 2.26$. So the 95% confidence interval for the population mean is

$$12.1 \pm \frac{2.26 \times 0.965}{\sqrt{10}} = 12.1 \pm 0.69$$

The 95% confidence interval for the mean moisture level at normal speed and normal temperature is 11.4% to 12.8%.

(b) The confidence intervals have considerable overlap and therefore there is no evidence to conclude that moisture levels at high speed/high temperature and normal speed/normal temperature are different.

(c) The experiment has been successful since it indicates that higher throughput can be obtained without affecting moisture. However, properties other than moisture are important in tobacco production and a more extensive evaluation should have been carried out.

2.3 (a) $n = 12$, $\bar{x} = 44.5$, $s = 8.83$, $t = 2.20$ with 11 degrees of freedom. 95% confidence interval for the mean:

$$\bar{x} \pm \frac{ts}{\sqrt{n}} = 44.5 \pm \frac{2.20 \times 8.83}{\sqrt{12}}$$

$$= 44.5 \pm 5.6$$

We are therefore 95% confident that the population mean lies between 38.9 and 50.1 g/kg.

(b)
$$n = \left(\frac{ts}{c}\right)^2 = \left(\frac{2.20 \times 8.83}{2.0}\right)^2$$

$$= 95 \text{ samples per delivery}$$

(c) Dr Small's suspicions are confirmed. There is a high sample-to-sample variability (RSD of around 20%) which will still have a large effect even with the present method of composite sampling. (In fact Dr Small assumes the uncertainty of the composite sample will be of a similar order to the confidence interval obtained above.) To reduce the uncertainty he can take more samples, but our calculations show that he needs around 100 to reduce the confidence interval to ± 2.0. This is clearly impractical. Technical improvements are probably the only way forward. Two improvements worthy of consideration are:

(i) Taking a sample later in the process when some mixing or blending has occurred.

(ii) Instituting a better sampling method such as taking core instead of probe samples.

2.4 (a)

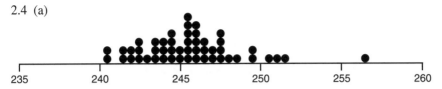

The observation for batch 16 has a much higher viscosity than the rest of the data. In practice a formal test of whether it was a statistical outlier would be carried out, plus an investigation of the process logs for the batch to see if a reason for the unusual result could be found before excluding it from the analysis.

(b)

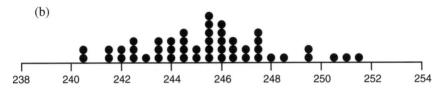

The distribution of blobs appears to be fairly symmetrical, with more values clustered around the centre than the extremes. There is some suggestion that the results spread out slightly at the high end, but more data would have to be collected to confirm this. If this proved to be the case then the decision to exclude the point at 256.3 would have to be re-examined.

(c) The population is that of the viscosities of all batches produced by the triphenolite process under its present operating conditions.

(d) Using the blob diagram we can see that, while no batches breached the lower specification, three were too high in viscosity and had to be scrapped. This gives a total cost of £3000. We can gain some idea of the likely cost savings if we displace the blob diagram down incrementally in 0.5 second steps (or move the limits up in 0.5 second steps).

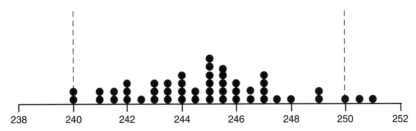

$\bar{x} = 245.0$: 0 low defects (£0), 2 high defects (£2000), total £2000.

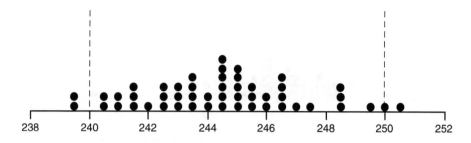

$\bar{x} = 244.5$: 2 low defects (£600), 1 high defect (£1000), total £1600.

Similarly

$\bar{x} = 244.0$: 2 low defects (£600), 0 high defects (£0), total £600
$\bar{x} = 243.5$: 4 low defects (£1200), 0 high defects (£0), total £1200
$\bar{x} = 243.0$: 6 low defects (£1800), 0 high defects (£0), total £1800

Although we could not expect another sample of 50 batches to reproduce the distribution exactly, the general shape should be fairly similar. A mean viscosity of around 244 seconds appears to minimise waste costs.

Chapter 3

3.1 Number of values $= 80$
Minimum COD $= 45.3$
Maximum COD $= 88.2$

A suitable grouping would be 45.0–49.9, 50.0–54.9, ...
This would give nine groups, which, for 80 values, is very suitable.
Tally count:

Class Interval	Tally	Frequency
45.0–49.9	ЖНҬ ЖНҬ ‖	12
50.0–54.9	ЖНҬ ЖНҬ ЖНҬ ‖‖‖	19
55.0–59.9	ЖНҬ ЖНҬ ЖНҬ ‖	16
60.0–64.9	ЖНҬ ЖНҬ ‖	11
65.0–69.9	ЖНҬ ‖‖‖	9
70.0–74.9	ЖНҬ	5
75.0–79.9	‖‖‖	4
80.0–84.9	‖‖	3
85.0–89.9	‖	1

This gives the following histogram:

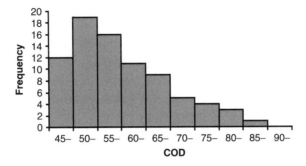

However, comparisons are difficult because of the larger number of values in the first histogram (before the unit was installed).

To make the comparison easier it is often useful to use percentage frequencies

$$\left(\frac{\text{Frequency}}{\text{Total no. of values}} \times 100 \right) \text{ as shown below.}$$

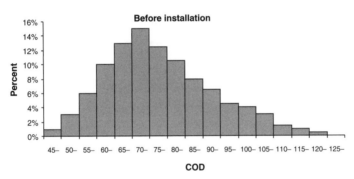

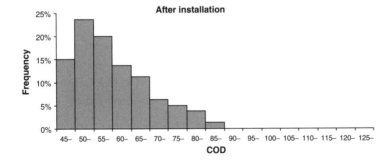

This clearly shows a reduction in frequency in the 70–90 mg/l range and no samples above 90 mg/l. However there is a greater frequency in the 45–60 mg/l range. It looks as if the unit has been very effective is reducing the high COD values (70–125 mg/l) but has little effect on lower COD values.

3.2 (a) Putting the data in order gives:

3, 6, 7, 11, 12, 14, 15, 15, 16, 17, 17, 19, 20, 22, 22, 24, 28, 35, 42

There are 19 observations.

   (i) The median is the 10th highest observation: 17.
  (ii) The lower quartile is the 5th lowest observation in magnitude: 12.
       The upper quartile is the 5th highest observation in magnitude: 22.
 (iii) The interquartile range = 22 − 12
                               = 10
  (iv) The lowest possible value of a whisker
             = lower quartile − (1½ × interquartile range)
             = 12 − (1½ × 10)
             = −3
       The lowest time was 3 and this will be the value of the low whisker.
       The highest possible value of a whisker = 22 + (1½ × 10)
                                               = 37
       The highest value between 22 and 37 is 35 and this is the value of
       the high whisker.
       There is one value beyond the whiskers, namely 42, and this will be
       drawn as an extreme value.
   (v) The whiskers are at 3 and 35.

(b)

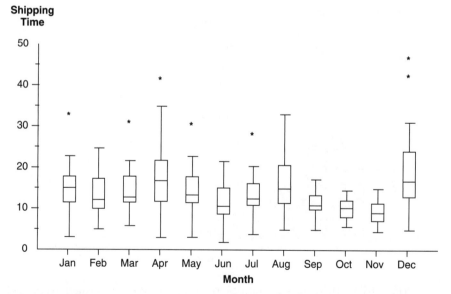

(c) Clearly there are differences in performance over the year. In December,
    April and August the performance is far poorer. This could be due to
    problems caused by extended holidays. In the period from September

to November the variability is much reduced – clearly there must be a message for Meg in this period.

Chapter 4

4.1 (a) **Null hypothesis**   The mean strength equals 5.4 ($\mu = 5.4$).

Alternative hypothesis   The mean strength does not equal 5.4 ($\mu \neq 5.4$).

The population refers to all batches produced **after** the modification. Notice that we do not refer at all to batches produced before the modification.

(b) $n = 10$, $\bar{x} = 4.5$, $s = 1.144$ with 9 degrees of freedom

Test value
$$= \frac{|\bar{x} - \mu|\sqrt{n}}{s}$$
$$= \frac{|4.5 - 5.4|\sqrt{10}}{1.144}$$
$$= 2.49$$

Table value   From Table A.2 with 9 degrees of freedom: 2.26 at the 5% significance level.

Decision   We reject the null hypothesis at the 5% significance level.

Conclusion   We can conclude that the mean strength is not 5.4 and therefore the modification has resulted in change of strength.

4.2 (a) **Null hypothesis**   The population mean is 50.0 ($\mu = 50.0$).

Alternative hypothesis   The population mean is not equal to 50.0 ($\mu \neq 50.0$).

If this is an initial evaluation it may well be carried out by a single highly skilled operator on the same day. The population for this situation is 'all possible values that could be obtained by the operator on that day'.

If, however, the method is shortly to become a standard procedure in the laboratory the population of interest is 'all possible values that could be obtained by all possible operators under all possible conditions'. The sample will have to be representative of this population.

(b) $n = 8$, $\bar{x} = 49.60$, $s = 0.968$ with 7 degrees of freedom

Test value
$$= \frac{|\bar{x} - \mu|\sqrt{n}}{s}$$
$$= \frac{|49.6 - 50.0|\sqrt{8}}{0.968} = 1.17$$

Table value   From Table A.2 with 7 degrees of freedom: 2.36 at the 5% significance level.

Decision   We cannot reject the null hypothesis.

Conclusion   There is no evidence that the method is biased.

(c) The 95% confidence interval is given by:

$$\bar{x} \pm \frac{ts}{\sqrt{n}} = 49.60 \pm \frac{2.36 \times 0.968}{\sqrt{8}}$$

$$= 49.60 \pm 0.81$$

i.e.   48.79 to 50.41 mg/litre

The maximum bias, at a 95% confidence level, is $(50 - 48.79) = 1.21$ mg/litre.

(d)
$$n = \left(\frac{ts}{c}\right)^2$$

$$= \left(\frac{2.36 \times 0.968}{0.5}\right)^2$$

$$= 21 \text{ determinations}$$

Thus 21 determinations are required to give a 95% confidence interval to within ± 0.5 mg/litre.

4.3 (a) The sample SD $(s) = 0.169$ with 5 degrees of freedom
From Table A.7 with 5 degrees of freedom:

$$k_a = 0.62; \quad k_b = 2.45$$

$$\text{Lower 95\% limit} = 0.62 \times 0.169 = 0.105$$

$$\text{Upper 95\% limit} = 2.45 \times 0.169 = 0.414$$

The 95% confidence interval for the population SD is 0.105 to 0.414.

(b) The 95% confidence interval for the SD of 0.105 to 0.414 includes the 'average' SD for all laboratories of 0.15. There is no evidence that the precision of Nulabs is different from that of the 'average' laboratory.

4.4 (a) $n_A = 10$; $SD_A = 2.951$

$n_B = 9$; $SD_B = 1.323$

Null hypothesis   The population standard deviations are equal.

Alternative hypothesis   The population standard deviations are unequal.

Test value $= \dfrac{SD_A^2}{SD_B^2}$

$= \dfrac{2.951^2}{1.323^2}$

$= 4.98$

Table value   With 9 degrees of freedom for the numerator and 8 for the denominator, the F-table, Table A.6 (two-sided), gives a value of 4.36 at the 5% significance level.

Decision   Since the test value is greater than the table value at a 5% significance level we reject the null hypothesis and accept the alternative.

Conclusion   From this evidence it would appear that method B is more repeatable.

(b) It must be emphasised that the population consists of determinations from **one** operator and **one** sample of cement and therefore the conclusion is of limited applicability.

4.5 (a) For Green's data, $n = 2$, $\bar{x} = 19.4$, $s = 0.283$ with 1 degree of freedom

Null hypothesis   The true mean net weight is equal to 20 tonnes ($\mu = 20$).

Alternative hypothesis   The true mean net weight is not equal to 20 tonnes ($\mu \neq 20$).

Test value $= \dfrac{|\bar{x} - \mu|\sqrt{n}}{s}$

$= \dfrac{|19.4 - 20|\sqrt{2}}{0.283}$

$= 3.00$

Table value   From Table A.2 with 1 degree of freedom: 12.71 at the 5% significance level.

Decision   We cannot reject the null hypothesis at the 5% significance level.

Conclusion   There is insufficient evidence to say that the mean net weight of gravel deliveries is different from 20 tonnes.

For Brown's data, $n = 7$, $\bar{x} = 20.1$, $s = 0.300$ with 6 degrees of freedom

Null hypothesis	The true mean net weight is equal to 20 tonnes ($\mu = 20$).
Alternative hypothesis	The true mean net weight is not equal to 20 tonnes ($\mu \neq 20$).

Test value
$$= \frac{|\bar{x} - \mu|\sqrt{n}}{s}$$
$$= \frac{|20.1 - 20|\sqrt{2}}{0.300}$$
$$= 0.88$$

Table value	From Table A.2 with 6 degrees of freedom: 2.45 at the 5% significance level.
Decision	We cannot reject the null hypothesis at the 5% significance level.
Conclusion	There is insufficient evidence to say that the mean net weight of gravel deliveries is different from 20 tonnes.

Both sets of data arrive at the same conclusion! Although Green's mean is below 20, it would be unwise to start making demands on the basis of only two data points. Brown's larger data set is more reassuring, but he should remember that the specification demands a mean of *at least* 20 tonnes. This test simply shows that from the available evidence we cannot show that the true mean is different from 20, but a value of less than 20 is not impossible. It is perhaps a little early to say 'there is nothing to worry about'.

(b) We cannot use the range to compare variability between samples of different sizes, as the range is dependent on sample size. Since the standard deviation is not dependent on sample size, the F-test is an appropriate significance test.

$s_{Green} = 0.283$ with 1 degree of freedom, $s_{Brown} = 0.300$ with 6 degrees of freedom

Null hypothesis	In the long run, the standard deviations of Green's and Brown's data will be the same ($\sigma_{Green} = \sigma_{Brown}$).
Alternative hypothesis	In the long run, the standard deviations of Green's and Brown's data will not be the same ($\sigma_{Green} \neq \sigma_{Brown}$).

Test value	$= \dfrac{\text{Larger SD}^2}{\text{Smaller SD}^2}$
	$= \dfrac{0.300^2}{0.283^2}$
	$= 1.12$

Table value   From Table A.6 (two-sided) with 6 and 1 degrees of freedom: 937.1 at 5% significance level.

Decision   We cannot reject the null hypothesis.

Conclusion   Brown's results cannot be ignored on the grounds of excessive variability.

The large table value of 937.1 should be a reminder to us of the difficulty of making reliable estimates about variability on the basis of very small data sets.

(c) For all nine results together, $\bar{x} = 19.94$, $s = 0.416$ with 8 degrees of freedom

95% confidence interval for the true mean

$$\bar{x} \pm \frac{ts}{\sqrt{n}} = 19.94 \pm \frac{2.31 \times 0.416}{\sqrt{9}}$$

$$= 19.94 \pm 0.32$$

$$= 19.62 \text{ to } 20.26 \text{ tonnes}$$

The best available estimate for the true mean net weight of gravel is 19.94 tonnes, marginally under the requirement that the true mean be at least 20 tonnes. The width of the confidence interval, however, indicates that the true value could still be as low as 19.62 (although of course it could be as high as 20.26). You would be well advised to discuss with the MD what margin of error he can live with and what amount of check weighing effort will be involved to deliver this to him.

As for his 44% statistic, he might just as well have said 56% of trucks are on target or overloaded! A perfectly targeted, perfectly symmetrical population of loads would have 50% below 20 tonnes. It is more pertinent to ask whether the load-to-load variability (reflected in a standard deviation of 0.416 tonnes) will cause any short-term supply shortages for Mixtup despite them receiving the right amount of gravel in the long run. Unless this is the case there does not appear to be any need to contact Quickhole.

Chapter 5

5.1 (a) Total area in both tails = 5%
From Table A.1 (two-tailed)
the value of z is 1.96.
Normal range:
$37.5 \pm (1.96 \times 0.3)$
i.e. 37.5 ± 0.59
i.e.   36.91% to 38.09%.
The Crunch output is given below.

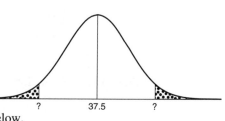

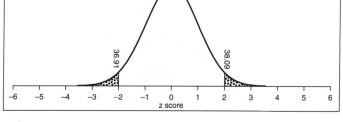

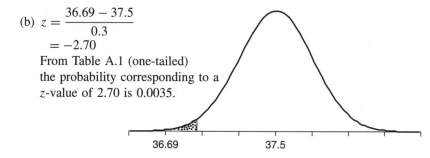

To obtain this output, enter everything you know in the shaded cells: Mean 37.5, SD 0.3.

The percentage of the distribution inside the limits is 95%, so there are 5% outside, or expressed as a fraction, 0.05. Split that equally between the two tails, 0.025 each.

In the unshaded cells the program calculates what you want to know – the limits are at 36.91 and 38.09.

(b) $z = \dfrac{36.69 - 37.5}{0.3}$
$= -2.70$
From Table A.1 (one-tailed)
the probability corresponding to a
z-value of 2.70 is 0.0035.

Table A.1 relates to a shaded area to the right of a positive z-value, but because of the symmetry of the normal distribution we can also use it for the negative tail.
(Clear the probabilities from the shaded cells, enter 36.69 as lower x limit.)
Thus we should expect a bottle with 36.69% strength or less to occur with probability 0.0035 or 0.35% of the time.

(c) A value of 36.69% is unusually low. Further samples would be needed to determine the situation, as the value might be an outlier, or the mean or standard deviation may have changed.

5.2 (a) The nominal weight is mid-way between the specification limits, i.e. the limits are symmetrical about the mean.

$z = (x - \mu)/\sigma$

$\quad = (230 - 200)/12$

$\quad = 2.5$

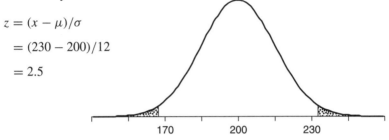

Using Table A.1 (two-tailed), Prob. = 0.0124 (1.24%)
The percentage of tablets outside specification is 1.24%.
(Mean 200, SD 12, lower x limit 170, upper x limit 230.)

(b) The nominal weight is not mid-way between the specification limits. The limits are not symmetrical about the mean.
We need to refer to Table A.1 (one-tailed) at each limit.
At the upper limit (230)

$z = (230 - 210)/12 = 1.67$

Prob. = 0.0475(4.75%)

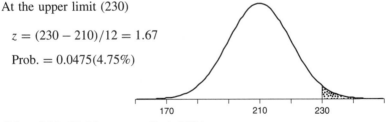

(Mean 210, SD 12, upper x limit 230.)
At the lower limit (170)

$$z = (170 - 210)/12 = -3.33$$

$$\text{Prob.} = 0.0004(0.04\%)$$

(Mean 210, SD 12, lower x limit 170.)
The percentage of tablets outside specification is

$$4.75 + 0.04 = 4.79\%$$

(c) If 99.9% are within the specification limits, 0.1% are in the tails. From the percentage points at the foot of Table A.1 (two-tailed)

$$z = 3.29$$

The limits are at

$$200 \pm (3.29 \times 12)$$

i.e. 200 ± 39.5

i.e.   160.5 to 239.5 μg

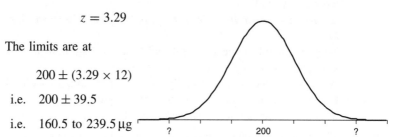

(Mean 200, SD 12, 0.1% = probability of 0.001 in the tails, divide equally: 0.0005 in each tail.)

(d) The limits are at

$$200 \pm (3.29 \times 8)$$

i.e. 200 ± 26.3

i.e.   173.7 to 226.3 μg

(Change SD to 8.)

(e) The mean is mid-way between the limits, i.e. the limits are symmetrical about the mean.

$$z = (x - \mu)/\sigma$$

so $\sigma = (x - \mu)/z$

$$= (230 - 200)/3.29$$

$$= 9.12 \, \mu g$$

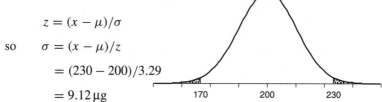

(Mean 200, SD unknown, lower x limit 170, upper x limit 230, probabilities 0.0005 in each tail.)

(f) Assuming the area below 170 is negligible, the area above 230 will be close to 0.1% (0.001). Using Table A.1 (one-tailed) gives:

$$z = 3.09$$

so $\sigma = (230 - 210)/3.09$

$$= 6.5 \, \mu g$$

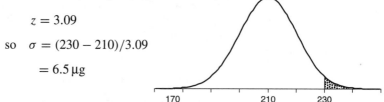

(Mean 210, SD unknown, lower x limit 170, upper x limit blank, probability all 0.001 in lower tail.)

(g) If we could set the population mean exactly at 200 µg, a population standard deviation of 9.1 µg would be satisfactory. However, this is unlikely and the population mean will be different from 200 µg. If it was 210 µg (or 190 µg) we would need a population standard deviation of 6.5 µg, but surely the mean could be controlled closer than 10 µg when the standard deviation is only 6.5 µg? A population standard deviation of around 8 µg might seem a reasonable compromise.

Chapter 8

8.1 The blob diagrams and summary statistics are as follows:

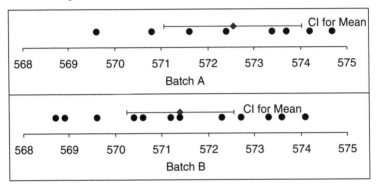

Batch A: $n = 8$, Mean = 572.55, SD = 1.776
Batch B: $n = 12$, Mean = 571.40, SD = 1.827

(a)

$$\text{Combined SD } (s) = \sqrt{\frac{(\text{df}_A \times \text{SD}_A^2) + (\text{df}_B \times \text{SD}_B^2)}{\text{df}_A + \text{df}_B}}$$

$$= \sqrt{\frac{(7 \times 1.776^2) + (11 \times 1.827^2)}{7 + 11}}$$

$$= 1.807 \text{ with 18 degrees of freedom}$$

(b) **Null hypothesis**   The populations mean melting points of the two batches are equal ($\mu_A = \mu_B$).

Alternative hypothesis   The populations mean melting points of the two batches are **not** equal ($\mu_A \neq \mu_B$).

Test value
$$= \frac{|\bar{x}_A - \bar{x}_B|}{s\sqrt{\dfrac{1}{n_A} + \dfrac{1}{n_B}}}$$

$$= \frac{|572.55 - 571.40|}{1.807\sqrt{\dfrac{1}{8} + \dfrac{1}{12}}}$$

$$= 1.39$$

Table value   From Table A.2 with 18 degrees of freedom: 2.10 at the 5% level of significance.

Decision   We cannot reject the null hypothesis.

Conclusion   There is insufficient evidence to conclude that the batches have different melting points.

(c) A confidence interval for the difference between the two population means is given by:

$$|\bar{x}_A - \bar{x}_B| \pm \left(ts\sqrt{\frac{1}{n_A} + \frac{1}{n_B}} \right)$$

$$= |572.55 - 571.40| \pm \left(2.10 \times 1.807\sqrt{\frac{1}{8} + \frac{1}{12}} \right)$$

$$= 1.15 \pm 1.73$$

$$= -0.58 \text{ to } +2.88\,°C$$

(d) Required sample sizes are given by:

$$n_A = n_B = 2\left(\frac{ts}{c} \right)^2$$

$$= 2 \times \left(\frac{2.10 \times 1.807}{1.2} \right)^2$$

$$= 20$$

We need approximately 20 samples from each batch to obtain a 95% confidence interval of $\pm 1.2\,°C$.

The *Crunch* output for this problem is as follows.

Mean	572.55	571.4
SD	1.776	1.827
Observations	8	12
df	7	11

Significance Level	5.0%

F-test for Standard Deviations	
Test Value	1.06
Table Value	4.71
Decision	Not Significant
P	0.977

Data Entry	
Batch A	Batch B
571.6	574.1
574.7	570.6
574.2	573.3
569.6	569.6
570.8	572.3
572.4	571.4
573.7	568.9
573.4	573.6
	568.7
	571.2
	572.7
	570.4

2-Sample t-test for Independent Samples	
Combined Standard Deviation	1.807
Test Value	1.39
Table Value	2.10
Decision	Not Significant
P	0.180

Confidence Interval for Difference between Means	
Confidence Level	95.0%
Least Significant Difference	1.733
Confidence Interval	−0.583
for True Difference	to
	2.883

Required Sample Size Calculation	
Required Difference to Estimate	1.2
Required Sample Size ($n_A = n_B$)	20.0
Total Sample Size ($n_A + n_B$)	40.0

8.2 (a) **Null hypothesis**   The population standard deviations of quarterly scores are equal.

Alternative hypothesis   The population standard deviations of quarterly scores are **not** equal.

Test value
$$= \frac{\text{Larger SD}^2}{\text{Smaller SD}^2}$$
$$= \frac{\text{2nd Quarter SD}^2}{\text{1st Quarter SD}^2}$$
$$= \frac{8.09^2}{5.39^2}$$
$$= 2.26$$

Table values   Using Table A.6 (two-sided) with 10 and 12 degrees of freedom gives:

3.37 at the 5% significance level.

Decision   We cannot reject the null hypothesis.

Conclusion   There is no evidence that the scores in the two quarters have different levels of variability.

We can now combine the standard deviations.

$$\text{Combined SD } (s) = \sqrt{\frac{(\text{df}_A \times \text{SD}_A^2) + (\text{df}_B \times \text{SD}_B^2)}{\text{df}_A + \text{df}_B}}$$

$$= \sqrt{\frac{(12 \times 5.39^2) + (10 \times 8.09^2)}{12 + 10}}$$

$$= 6.75 \text{ with } 12 + 10 = 22 \text{ degrees of freedom}$$

(b) **Null hypothesis** — The population mean quarterly scores are equal ($\mu_A = \mu_B$).

Alternative hypothesis — The population mean quarterly scores are **not** equal ($\mu_A \neq \mu_B$).

Test value

$$= \frac{|\bar{x}_A - \bar{x}_B|}{s\sqrt{\dfrac{1}{n_A} + \dfrac{1}{n_B}}}$$

$$= \frac{|84.23 - 78.09|}{6.75\sqrt{\dfrac{1}{13} + \dfrac{1}{11}}}$$

$$= 2.22$$

Table value — From Table A.2 with 22 degrees of freedom:

2.07 at the 5% significance level.

(The value for 22 degrees of freedom is not given in Table A.2 so it is necessary to interpolate between the values for 20 and 25 degrees of freedom.)

Decision — Reject the null hypothesis.

Conclusion — Customer satisfaction has dropped from the first quarter to the second quarter.

(c) The 95% confidence interval for the difference between the means of the two quarters is:

$$|\bar{x}_A - \bar{x}_B| \pm ts\sqrt{\frac{1}{n_A} + \frac{1}{n_B}} = |84.23 - 78.09| \pm 2.07 \times 6.75\sqrt{\frac{1}{13} + \frac{1}{11}}$$

$$= 6.14 \pm 5.73$$

$$= 0.4 \text{ to } 11.9$$

Thus we can be 95% certain that the mean customer satisfaction score in the first quarter was higher than in the second quarter by between 0.4 and 11.9.

(d) The sample sizes required to give a 95% confidence interval of ± 3.0 are:

$$n_A = n_B = 2 \left(\frac{ts}{c} \right)^2$$

$$= 2 \times \left(\frac{2.07 \times 6.75}{3.0} \right)^2$$

$$= 44 \text{ retailers}$$

Thus 44 replies would be needed in each quarter.
Here is the *Crunch* output for this problem.

Mean	84.23	78.09
SD	5.388	8.093
Observations	13	11
df	12	10

Significance Level	5.0%

F-test for Standard Deviations	
Test Value	2.26
Table Value	3.37
Decision	Not Significant
P	0.183

Data Entry	
1st Quarter	2nd Quarter
85	69
81	82
88	92
92	76
76	78
78	83
93	87
81	67
79	73
84	69
86	83
82	
90	

2-Sample t-test for Independent Samples	
Combined Standard Deviation	6.753
Test Value	2.22
Table Value	2.07
Decision	Significant
P	0.037

Confidence Interval for Difference between Means	
Confidence Level	95.0%
Least Significant Difference	5.737
Confidence Interval	0.403
for True Difference	to
	11.877

Required Sample Size Calculation	
Required Difference to Estimate	3
Required Sample Size ($n_A = n_B$)	43.6
Total Sample Size ($n_A + n_B$)	87.2

Chapter 9

9.1 (a) We use the paired t-test.
The increase in number of miles per gallon for each car when using the additive is:

Car	A	B	C	D	E	F
Increase in mpg	1.6	1.1	1.7	0.3	0.3	2.2

$$\text{Mean difference } (\bar{x}_d) = 1.2$$

SD of differences $(s_d) = 0.780$ with 5 degrees of freedom

Null hypothesis $\quad \mu_d = 0$

Alternative hypothesis $\quad \mu_d \neq 0$

Test value
$$= \frac{|\bar{x}_d - \mu_d|\sqrt{n_d}}{s_d}$$
$$= \frac{|1.2 - 0|\sqrt{6}}{0.780}$$
$$= 3.77$$

Table values — From Table A.2 with 5 degrees of freedom: 2.57 and 4.03 at the 5% and 1% levels, respectively.

Decision — We reject the null hypothesis at a 5% significance level.

Conclusion — There is substantial evidence of a decrease in petrol consumption.

(b) We have $s_d = 0.780$ with 5 degrees of freedom, $c = 1.0, t = 2.57$. So

$$n_d = \left(\frac{ts_d}{c}\right)^2$$
$$= \left(\frac{2.57 \times 0.780}{1.0}\right)^2$$
$$= 4.0$$

A sample size of 4 cars is needed to be certain of obtaining a 95% confidence interval within $\pm 1.0\,\text{m.p.g.}$

Here is the *Crunch* **output** for this problem so far.

With	Without	Differences
26.2	24.6	1.6
35.1	34	1.1
43.2	41.5	1.7
36.2	35.9	0.3
29.4	29.1	0.3
37.5	35.3	2.2

CI for Mean

| 0 | 0.5 | 1 | 1.5 | 2 | 2.5 |

Differences

Summary Statistics for Differences

Mean	1.2
SD	0.78
Count	6
df	5

Adjust Rounding

Paired t-test

Significance Level	5.0%
Reference Value	0
Test Value	3.77
Table Value	2.57
Decision	Significant
P	0.013

(c) The standard deviations would be:

With additive : $s_A = 6.05$

Without additive : $s_W = 5.86$

Combined SD : $s = 5.95$ (10 degrees of freedom)

$$n_A = n_W = 2\left(\frac{ts}{c}\right)^2$$

$$= 2\left(\frac{2.23 \times 5.95}{1.0}\right)^2$$

$$= 352$$

Thus 704 cars would be required, compared with only four for the paired design.

9.2 The relevant data required for the statistical tests, after extracting and tabulating in relation to formulations, is as follows:

Panellist	A	B	Diff		Panellist	C	D
1	31	34	−3		2	37	39
3	58	58	0		5	39	53
4	56	51	5		8	30	39
6	58	47	11		9	48	59
7	57	59	−2		11	44	58
10	35	30	5		12	61	79
13	60	46	14		14	53	62
16	49	37	12		15	30	36
18	47	48	−1		17	50	60
20	36	29	7		19	33	52
21	59	60	−1		22	53	66
24	52	47	5		23	32	43
Mean	49.8	45.5	4.3		Mean	42.5	53.8
SD	10.4	10.9	5.8		SD	10.5	12.8

(a) Since A and B were applied to the same panellist, we use the paired t-test.

Null hypothesis There is no difference in the effectiveness of A and B (the mean of the population of differences equals zero, $\mu_d = 0$).

Alternative hypothesis There is a difference in the effectiveness of A and B (the mean of the population of differences does **not** equal zero, $\mu_d \neq 0$).

Test value
$$= \frac{|\bar{x}_d - \mu_d|\sqrt{n_d}}{s_d}$$

$$= \frac{|4.3 - 0|\sqrt{12}}{5.8} = 2.57$$

where $\bar{x}_d$ is the sample mean of the differences;

μ_d the mean difference;

s_d the sample standard deviation of differences.

Table value From Table A.2 with 11 degrees of freedom: 2.20 at 5% significance level.

Decision Reject the null hypothesis at 5% significance level.

Conclusion We conclude that the formulations A and B are not equally effective, and that B is more effective as its mean value is lower.

(b) The required number of panellists for a mean difference of 5.0 is given by

$$\left(\frac{t_d s_d}{c}\right)^2 = \left(\frac{2.20 \times 5.8}{5.0}\right)^2 = 6.6$$

A panel of at least 7 would be required.

(c) To compare B and D we must notice that they were applied to two different groups of panellists. We need to use the two-sample t-test for independent samples.

Null hypothesis There is no difference in the effectiveness of formulations B and D ($\mu_B = \mu_D$).

Alternative hypothesis There is a difference in the effectiveness of formulations B and D ($\mu_B \neq \mu_D$).

Combined standard deviation $= \sqrt{\dfrac{\left(\mathrm{df_B} \times \mathrm{SD_B^2}\right) + \left(\mathrm{df_D} \times \mathrm{SD_D^2}\right)}{\mathrm{df_B} + \mathrm{df_D}}}$

$$= \sqrt{\frac{(11 \times 10.9^2) + (11 \times 12.8^2)}{11 + 11}}$$

$= 11.9$ with 22 degrees of freedom

Test value $= \dfrac{|\bar{x}_B - \bar{x}_D|}{s\sqrt{\dfrac{1}{n_B} + \dfrac{1}{n_D}}}$

$$= \frac{|45.5 - 53.8|}{11.9\sqrt{\dfrac{1}{12} + \dfrac{1}{12}}} = 1.71$$

Table value From Table A.2 with 22 degrees of freedom:
2.07 at 5% significance level.

Decision We cannot reject the null hypothesis.

Conclusion We have insufficient evidence that there is a difference in effectiveness between formulations B and D.

(d) Required number of panellists **for each formulation** for a difference of 5.0 is given by

$$2\left(\frac{ts}{c}\right)^2 = 2\left(\frac{2.07 \times 11.9}{5.0}\right) = 48.7$$

Approximately 50 panellists for each formulation (100 in all) would be required.

(e) Where the experiment allowed A and B to be compared directly and the test was **paired**, a mean difference of 4.3 was statistically significant.

To compare B and D involved different groups of panellists. A difference between the means of 8.3 was not significant.

Where pairing is possible, the experiment is more sensitive and the test more powerful.

Notice how pairing eliminates much of the inherent variability in this type of investigation – person-to-person. The standard deviation of readings for one formulation was over 10, but the differences had a standard deviation of only 5, showing one of the benefits of pairing.

Another consequence is the much smaller panel size needed in a paired experiment to detect a particular difference – for a difference of 5.0 to be significant, a panel of around 100 would be needed for indirect comparison in the two-sample t-test (50 for each formulation) but only 7 panellists would be required for a paired experiment.

(f) Although it would not be possible to compare four formulations simultaneously on the same person, the experiment could be done over two days and the same panel used throughout.

Chapter 10

10.1 (a) The Instant Count readings have been put on the x-axis, so must be the independent variable. The log SPC must therefore be the response. They could equally well have been chosen the other way round, but this choice was made as we wish to predict log SPC from an Instant Count reading.

(b)

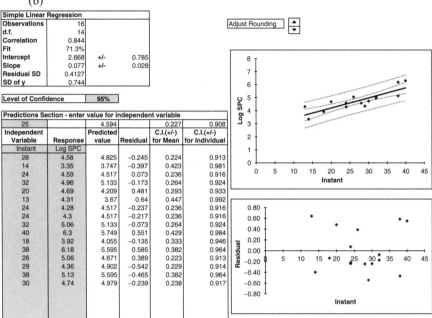

Simple Linear Regression			
Observations	16		
d.f.	14		
Correlation	0.844		
Fit	71.3%		
Intercept	2.668	+/-	0.785
Slope	0.077	+/-	0.028
Residual SD	0.4127		
SD of y	0.744		

Level of Confidence	95%

Predictions Section - enter value for independent variable				
25		4.594		0.227	0.908
Independent Variable	Response	Predicted value	Residual	C.I.(+/-) for Mean	C.I.(+/-) for Individual
Instant	Log SPC				
28	4.58	4.825	-0.245	0.224	0.913
14	3.35	3.747	-0.397	0.423	0.981
24	4.59	4.517	0.073	0.236	0.916
32	4.96	5.133	-0.173	0.264	0.924
20	4.69	4.209	0.481	0.293	0.933
13	4.31	3.67	0.64	0.447	0.992
24	4.28	4.517	-0.237	0.236	0.916
24	4.3	4.517	-0.217	0.236	0.916
32	5.06	5.133	-0.073	0.264	0.924
40	6.3	5.749	0.551	0.429	0.984
18	3.92	4.055	-0.135	0.333	0.946
38	6.18	5.595	0.585	0.382	0.964
26	5.06	4.671	0.389	0.223	0.913
29	4.36	4.902	-0.542	0.229	0.914
38	5.13	5.595	-0.465	0.382	0.964
30	4.74	4.979	-0.239	0.238	0.917

The graph of the data suggests a linear relationship. The residuals are randomly scattered with no obvious pattern.

(c) The slope (b) is 0.077 (± 0.028)

The intercept (a) is 2.67 (± 0.78)

The best fit line is log SPC $= 2.67 + 0.077$ Instant

(d) Residual standard deviation

$$\text{RSD} = \sqrt{\frac{\text{Sum of the squared residuals}}{\text{Degrees of freedom}(= n - 2)}}$$

$$= \sqrt{\frac{0.245^2 + \ldots + 0.239^2}{14}}$$

$$= 0.41$$

(e) **Null hypothesis**   There is no linear association between Instant Count readings and log SPC.

    Alternative hypothesis   There is a linear association between the Instant Count readings and log SPC.

    Test value $|r| = 0.844$

    Table value   From Table A.12 with 14 degrees of freedom:

    0.497 at the 5% significance level;

    0.623 at the 1% significance level.

    Decision   We can reject the null hypothesis at the 1% significance level.

    Conclusion   There is a linear association between Instant Count readings and log SPC.

(f)

$$\text{Predicted log SPC} = a + bX$$

$$= 2.67 + 0.077 \times 25$$

$$= 4.59$$

95% confidence interval:

$$(a + bX) \pm \left(t \times \text{RSD}\sqrt{1 + \frac{1}{n} + \frac{(X - \bar{x})^2}{(n - 1)(\text{SD of } x)^2}} \right)$$

$$= 4.59 \pm \left(2.14 \times 0.41\sqrt{1 + \frac{1}{16} + \frac{(25 - 26.9)^2}{15 \times 8.16^2}} \right)$$

$$= 4.59 \pm 0.91$$

We are therefore 95% confident that the log SPC for that sample is between 3.68 and 5.50.

Converting to antilogs, we are 95% confident that the standard plate count is between 4.8×10^3 and 3.2×10^5 cfu/ml.

10.2 (a) (i) Blood lead is the response, water lead is the independent variable.

      (ii) There are no unusual patterns in the scatter diagram or plot of residuals.

      (iii) The test value is 0.803, the magnitude of the correlation coefficient.

The table values, with 10 degrees of freedom, are:

0.576 at the 5% significance level;

0.708 at the 1% significance level.

There is a significant relationship at the 1% level between water lead and blood lead.

(iv) With water lead at 160 μg/l, the predicted mean blood lead is 27.5 ± 5.1 μg/l.

The Crunch output for Problem 9.2(a) is as follows.

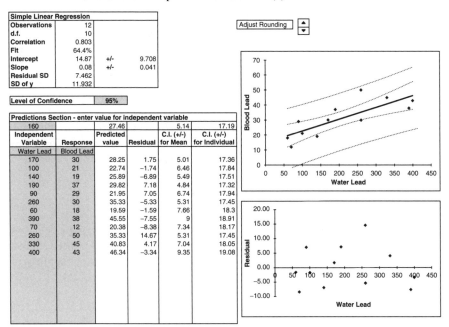

Simple Linear Regression			
Observations	12		
d.f.	10		
Correlation	0.803		
Fit	64.4%		
Intercept	14.87	+/-	9.708
Slope	0.08	+/-	0.041
Residual SD	7.462		
SD of y	11.932		

Level of Confidence	95%

Predictions Section - enter value for independent variable				
160		27.46		5.14	17.19
Independent		Predicted		C.I. (+/-)	C.I. (+/-)
Variable	Response	value	Residual	for Mean	for Individual
Water Lead	Blood Lead				
170	30	28.25	1.75	5.01	17.36
100	21	22.74	-1.74	6.46	17.84
140	19	25.89	-6.89	5.49	17.51
190	37	29.82	7.18	4.84	17.32
90	29	21.95	7.05	6.74	17.94
260	30	35.33	-5.33	5.31	17.45
60	18	19.59	-1.59	7.66	18.3
390	38	45.55	-7.55	9	18.91
70	12	20.38	-8.38	7.34	18.17
260	50	35.33	14.67	5.31	17.45
330	45	40.83	4.17	7.04	18.05
400	43	46.34	-3.34	9.35	19.08

(b) With years of occupancy as the independent variable:

(i) The scatter diagram is dominated by four points which lie well beyond the other points. Therefore any conclusions will be highly dependent upon these points.

(ii) The correlation coefficient is 0.390. Therefore there is no significant relationship.

(c) With age of occupant as the independent variable:

(i) There are no unusual patterns.

(ii) The correlation coefficient is 0.284. Therefore there is no significant relationship.

(d) (i) The question is a long-winded way of asking 'What is the slope?' since the slope is the unit increase in y per unit increase in x.

The slope is 0.079 ± 0.041. In other words, we can expect an increase in blood lead of between 0.038 and 0.120 µg/l for every 1 µg/l increase in water lead.

(ii) The width of the confidence interval depends upon the closeness of the value of the independent variable to $\bar{x}$ which has a value of 205. House D is closest to the mean and has the smallest width, and house L is the furthest away and thus has the highest width of confidence interval.

(iii) The correlation coefficient of the residuals versus years of occupancy is 0.811. Clearly there is a relationship. Thus we can conclude that blood lead is dependent upon water lead and years of occupancy – this **multiple regression** relationship has a percentage fit of 88%, a large improvement on the 64% of the **simple** regression of blood lead on water lead. It makes sense that the level of lead in the blood should depend on the level in the water and the time exposed to it.

Age of occupant had a correlation coefficient of 0.621 with the residuals. Although significant, it is not so strong and is partly caused by the correlation between age of occupant and years of occupancy.

Chapter 11

11.1 (a) The sample size is finite (20), the probability of any jar having a defective seal may be assumed to be the same for all jars, and the batch size (number of jars in each delivery) is large, so the binomial distribution applies.

(b) From Table A.10, with 1% defectives and a sample size of 20,

Probability of no defectives in sample = 0.818.

The probability of one or more is

$$1 - 0.818 = 0.182.$$

(c) A delivery is rejected if there is one defective or more within the sample. We have just obtained in (b) the probability of this occurring. The probability of rejecting a delivery containing 1% defectives is therefore 0.182, or only 18%.

(d) From Table A.10:

Sample size	Probability of no defectives in sample	Probability of rejecting delivery containing 1% defective
5	0.951	0.049
10	0.904	0.096
20	0.818	0.182
30	0.740	0.260
40	0.669	0.331
50	0.605	0.395
80	0.448	0.552
100	0.366	0.634

For the probability to be greater than 0.50, a sample size of 80 would be required. (The actual number is at least 69.)

11.2 (a)

Single Sampling Scheme	A	B	Upper limit for graph(%)	10		Single Sampling Scheme	
Sample Size	100	100				Percent	Ac = 3	Ac = 4
Acceptance No.	3	4				0.0	1.000	1.000

Percent	Ac = 3	Ac = 4
0.0	1.000	1.000
0.5	0.998	1.000
1.0	0.982	0.997
1.5	0.936	0.982
2.0	0.859	0.949
2.5	0.759	0.894
3.0	0.647	0.818
3.5	0.535	0.727
4.0	0.429	0.629
4.5	0.336	0.530
5.0	0.258	0.436
5.5	0.194	0.351
6.0	0.143	0.277
6.5	0.104	0.214
7.0	0.074	0.163
7.5	0.053	0.122
8.0	0.037	0.090
8.5	0.025	0.066
9.0	0.017	0.047
9.5	0.012	0.034
10.0	0.008	0.024

For an acceptance number of 3:
Probability of accepting a good batch (3% defectives) = 0.647
Probability of rejecting a good batch (3% defectives) = 1 − 0.647 = 0.353
Probability of accepting a bad batch (5% defectives) = 0.258
For an acceptance number of 4:
Probability of accepting a good batch (3% defectives) = 0.818
Probability of rejecting a good batch (3% defectives) = 1 − 0.818 = 0.182
Probability of accepting a bad batch (5% defectives) = 0.436
Neither scheme is satisfactory. In both the risks of accepting bad batches and of rejecting good batches are too great.

Single Sampling Scheme	A	B
Sample Size	500	600
Acceptance No.	19	23

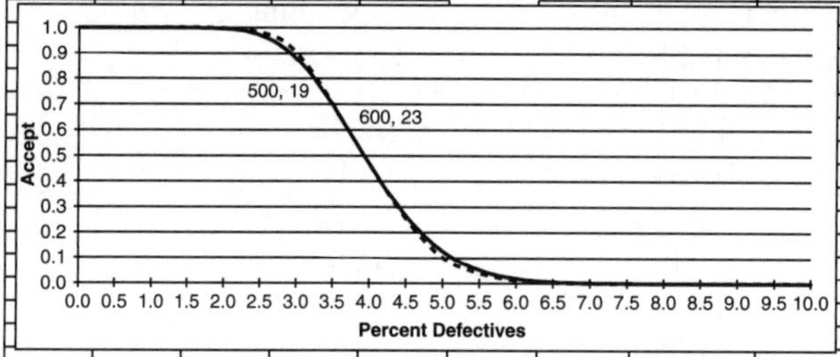

Upper limit for graph(%)   10

Single Sampling Scheme		
Percent	500,19	600, 23
0.0	1.000	1.000
0.5	1.000	1.000
1.0	1.000	1.000
1.5	1.000	1.000
2.0	0.997	0.999
2.5	0.971	0.982
3.0	0.879	0.902
3.5	0.697	0.719
4.0	0.468	0.471
4.5	0.265	0.250
5.0	0.127	0.109
5.5	0.053	0.039
6.0	0.019	0.012
6.5	0.006	0.003
7.0	0.002	0.001
7.5	0.000	0.000
8.0	0.000	0.000
8.5	0.000	0.000
9.0	0.000	0.000
9.5	0.000	0.000
10.0	0.000	0.000

Graph: Accept (y-axis, 0.0 to 1.0) vs Percent Defectives (x-axis, 0.0 to 10.0), showing curves labelled "500, 19" and "600, 23".

(b) The output shows two possible schemes.
The first uses a sample size of 500 and an acceptance number of 19.
This gives a probability of rejecting a good batch (3% defectives) of $1 - 0.879 = 0.121$ and a probability of accepting a bad batch (5% defectives) of 0.127.
The probabilities are of a similar magnitude but the sample size needs to be increased, keeping the acceptance number as a similar proportion of the sample size.
The second scheme uses a sample size of 600 and the acceptance number has been increased to 23.
This gives the probability of rejecting a good batch as $1 - 0.902 = 0.098$ and a probability of accepting a bad batch of 0.109.
These probabilities are similar and both close to 0.1. It is a very suitable scheme.

Chapter 12

12.1 (a)

$$\bar{x} = \frac{228}{60} = 3.8$$

The 95% confidence interval for the population mean is given by

$$\bar{x} \pm z\sqrt{\frac{\bar{x}}{n}}$$

From Table A.1 (two-tailed), $z = 1.96$. So

$$\bar{x} \pm z\sqrt{\frac{\bar{x}}{n}} = 3.8 \pm 1.96\sqrt{\frac{3.8}{60}}$$

$$= 3.8 \pm 0.49$$

$$= 3.31 \text{ to } 4.29$$

There is no evidence that the true mean is different from 4.0.
The assumption being made is that the number of strawberries per tart is a random event which follows a Poisson distribution with mean 3.8 (and hence standard deviation $= \sqrt{3.8}$).

(b)

No. of strawberries	Observed frequency	Poission probability	Expected frequency
0	2	0.0183	1.6
1	7	0.0733	6.6
2	11	0.1465	13.2
3	16	0.1954	17.6
4	21	0.1954	17.6
5	13	0.1563	14.1
6	11	0.1042	9.4
7	5	0.0595	5.4
8	3	0.0298	2.7
9	1	0.0132	1.2
10 or more	0	0.0081	0.7

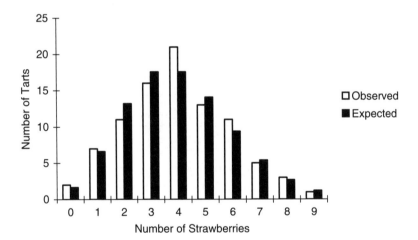

(c) The Poisson distribution provides a good fit to the data. Erdberry can inform his managing director that the strawberries are randomly distributed and no amount of mixing will improve the situation.

(d) Using *Crunch* and altering the mean gives a suitable value of 6.64.

Observed Count	Observed Frequency No. of Tarts	Poisson Probability
0	2	0.0183
1	7	0.0733
2	11	0.1465
3	16	0.1954
4	21	0.1954
5	13	0.1563
6	11	0.1042
7	5	0.0595
8	3	0.0298
9	1	0.0132
10	0	0.0053
Total Count	360	
Number of Observations	90	0.9972
Mean Count	4.00	4.00

Observed Count	Observed Frequency No. of Tarts	Poisson Probability
0	2	0.0025
1	7	0.0149
2	11	0.0446
3	16	0.0892
4	21	0.1339
5	13	0.1606
6	11	0.1606
7	5	0.1377
8	3	0.1033
9	1	0.0688
10	0	0.0413
Total Count	360	
Number of Observations	90	0.9574
Mean Count	4.00	6.00

With a mean of 4.00,

% defectives $= 1.83\% + 7.33\%$

$= 9.16\%$

With a mean of 6.00,

% defectives $= 0.25\% + 1.49\%$

$= 1.74\%$

Observed Count	Observed Frequency No. of Tarts	Poisson Probability
0	2	0.0013
1	7	0.0087
2	11	0.0288
3	16	0.0638
4	21	0.1059
5	13	0.1406
6	11	0.1556
7	5	0.1476
8	3	0.1225
9	1	0.0904
10	0	0.0600
Total Count	360	
Number of Observations	90	0.9251
Mean Count	4.00	6.64

With a mean of 6.64,

% defectives $= 0.13\% + 0.87\% = 1.00\%$

Probability of no strawberries $= 0.0013$

Probability of one strawberry $= \underline{0.0087}$

Probability of a defective $0.0100(1\%)$

12.2 (a) The mean number of jams per day is

$$\bar{x} = \frac{93}{30} = 3.1$$

The 95% confidence interval for the mean is

$$\bar{x} \pm 1.96\sqrt{\frac{\bar{x}}{n}} = 3.1 \pm 1.96\sqrt{\frac{3.1}{30}}$$

$$= 3.1 \pm 0.63$$

$$= 2.47 \text{ to } 3.73 \text{ jams per day}$$

(b)

Before : 93 jams in 30 days = 3.1 jams per day

After : 41 jams in 20 days = 2.05 jams per day

Null hypothesis In the long run there is no difference in the number of jams before and after the introduction of high quality paper.

Alternative hypothesis There is a difference.

Test value
$$= \frac{\left(\dfrac{\text{Count}_A}{n_A} - \dfrac{\text{Count}_B}{n_B}\right)}{\sqrt{\dfrac{\text{Count}_A + \text{Count}_B}{n_A \times n_B}}}$$

$$= \frac{\dfrac{93}{30} - \dfrac{41}{20}}{\sqrt{\dfrac{93 + 41}{30 \times 20}}} = 2.22$$

Table value From Table A.1:
1.96 at a 5% significance level;
2.58 at a 1% significance level.

Decision Reject the null hypothesis at a 5% significance level.

Conclusion There has been an improvement. However, the improvement could be due to other variables rather than the change in paper.

Chapter 13

13.1

Grade \ Time	Before		After		Total
Hazed	40	77%	12	23%	52
	8%		4%		7%
Not hazed	460	61%	288	39%	748
	92%		96%		94%
Total	500	63%	300	38%	800

Test Value	2.22
Significance Level	5%
Table value	1.96
Decision	Significant

Null hypothesis The percentage of hazed insulators is the same before and after the modification.

Alternative hypothesis The percentage of hazed insulators changed after the modification.

Test value
$$= \frac{|P_1 - P_2|}{\sqrt{\bar{P}(100 - \bar{P})\left(\dfrac{1}{n_1} + \dfrac{1}{n_2}\right)}} = 2.22$$

Table value From Table A.1 (two-tailed): 1.96 at the 5% significance level.

Decision Reject the null hypothesis.

Conclusion The percentage of hazed insulators changed after the modification.

Before the modification it was 8%, afterwards it was reduced to 4%.

13.2 (a)

Grade \ Level	Top		Upper Middle		Lower Middle		Bottom			Row Total
Mouldy	11	12.5	9	12.7	16	12.4	13	11.4		49
Slaty	5	10.7	10	10.9	8	10.6	19	9.8		42
First Class	473	465.9	481	476.3	461	462.0	417	427.8		1832
										0
										0
Column Total	489		500		485		449		0	1923

Test Value	15.33
df	6
Significance Level	5%

Table Value	12.59

Decision	Significant

Null hypothesis   There is no difference in the population percentage of mouldy or slaty beans from the four levels.

Alternative hypothesis   There are differences in the population percentage of mouldy or slaty beans from the four levels.

Test value $= \sum \dfrac{(\text{Observed} - \text{Expected})^2}{\text{Expected}} = 15.33$

Table value   From Table A.8 (one-sided) with $(3 - 1)$ $(4 - 1) = 6$ degrees of freedom: 12.59 at the 5% significance level.

Decision   We can reject the null hypothesis.

Conclusion   There are differences in population percentages at the four levels. The consignment is not homogeneous.

Grade \ Level	Top		Upper Middle		Lower Middle		Bottom			Row Total
Slaty	5	10.7	10	11.0	8	10.5	19	9.8		42
First Class	473	467.3	481	480.0	461	458.5	417	426.2		1832
										0
										0
										0
Column Total	478		491		469		436		0	1874

Test Value	12.74
df	3
Significance Level	5%

Least Significant Difference	n_1	468
between percentages	n_2	468
(data in 1st two rows only)	sig. level	5%

Table Value	7.81

5% LSD	1.90%

Decision	Significant

(b) Using only slaty and first class, there are significant differences in percentage slaty between the four levels.

The test value is 10.99, the table value with 3 degrees of freedom is 7.81.

The percentages of slaty for each level are:

Top	1.05%
Upper middle	2.04%
Lower middle	1.71%
Bottom	4.36%

Using average sample sizes of 468, the 5% least significant difference between percentages is 1.90%.

The percentages can be reported in order of magnitude, with those not significantly different bracketed together, as follows:

Top	1.05% ⎤
Lower middle	1.71% ⎥
Upper middle	2.04% ⎦
Bottom	4.36%

The table shows that there is no significant difference between the top three levels, but the bottom level is significantly different from all the other levels.

Cocoanuts believe that immature slaty beans are being hidden beneath good beans. This will make it very difficult to devise a suitable sampling inspection scheme.

Chapter 14

14.1 The sums of the ranks for the formulations are 47 for A and 89 for B.

Null hypothesis	The two formulations have the same degree of penetration.
Alternative hypothesis	The two formulations have different degrees of penetration.
Test value	Since there were equal numbers of observations, the test value is the larger of the two rank sums = 89.
Table value	From Table A.14 with sample sizes of 8 and 8:
	$86\frac{1}{2}$ at the 5% significance level,
	$92\frac{1}{2}$ at the 1% significance level.
Decision	We can reject the null hypothesis at the 5% level.
Conclusion	Formulation A has better penetration.

14.2 The scores are first converted to ranks, starting with the lowest score, which corresponds to the smallest difference between shampoos.

Subject	Score	Rank	More effective shampoo	Ranks for E	Ranks for F
2	3	1	F		1
4	5	2	E	2	
6	7	3	F		3
1	8	4	F		4
11	9	5	E	5	
7	10	6	E	6	
8	11	7.5	F		7.5
9	11	7.5	E	7.5	
12	12	9	E	9	
3	15	10	E	10	
10	16	11	E	11	
5	18	12	E	12	
		Totals		62.5	15.5

Null hypothesis	The two shampoos are equally effective in treating dandruff.
Alternative hypothesis	The two shampoos are not equally effective in treating dandruff.
Test value	The larger of the two rank totals = 62.5
Table value	$64\frac{1}{2}$ at the 5% significance level from Table A.15 with a sample size of 12.
Decision	We cannot reject the null hypothesis.
Conclusion	We are unable to conclude that the shampoos have different effectiveness in treating dandruff.

This conclusion does not 'prove' that Fledgeling's shampoo is as good as the market leader. The rankings give a strong suspicion that it is not, and the lack of a significant difference may be due to there being too small a panel for this type of investigation.

Chapter 15

15.1 (a)

Source of Variability	Sum of Squares	Degrees of Freedom	Mean Square
Between-days	44.62	7	6.37
Within-days	12.49	16	0.78
Total	57.11	23	

(b) Day 4 has the highest mean, 35.8.

(c) SD for day 4 = 1.04

(d) The p-value is shown as 0.000 to 3 decimal places so it is less than 0.001 (actually 0.0003). Between-days variability is easily significant at the 5% level.

(e) Within-days SD

$$= \sqrt{\text{Within-days mean square}}$$

$$= \sqrt{0.780}$$

$$= 0.88$$

Between-days SD

$$= \sqrt{\frac{\text{Between-days mean square} - \text{Within-days mean square}}{\text{Number of tests per day}}}$$

$$= \sqrt{\frac{6.374 - 0.780}{3}}$$

$$= 1.37$$

(f) Overall SD

$$= \sqrt{1.37^2 + 0.88^2}$$

$$= 1.63$$

(g) The major source of variability is between-days, which is the major contributor to the overall SD. To reduce variability MIRROR should concentrate on examining the daily setting-up conditions such as calibration.

Here is the *Crunch* output for Problem 15.1.

Significance Level	5.0%

One - Way Analysis of Variance Table							
Source	Sum of Squares	df	Mean Square	SD	F	Table Value	P
Between Days	44.62	7	6.37	1.37	8.17	2.66	0.000
Within Days	12.49	16	0.78	0.88			
Total	57.11	23		1.63			

Adjust Rounding	▲▼		▲▼	▲▼			

Adjust Rounding	▲▼						
Mean	32.7	34.2	32	35.8	31.7	33.8	34.5	35.1
Standard Deviation	0.91	1.08	0.90	1.04	0.58	0.62	1.15	0.55
Observations	3	3	3	3	3	3	3	3

Data Entry (upto 40 batches) >>>

Day

	Day 1	Day 2	Day 3	Day 4	Day 5	Day 6	Day 7	Day 8
Test	32.31	35.39	31.52	34.63	31.19	33.32	33.24	34.50
	33.79	33.24	33.05	36.67	32.33	33.57	35.50	35.13
	32.13	34.11	31.46	35.98	31.56	34.50	34.75	35.60

15.2 (a)

Source of Variability	Sum of Squares	Degrees of Freedom	Mean Square
Due to samples	2805.8	4	701.44
Due to testing	724.0	12	60.33
Total	3529.8	16	

(b) Test-to-test SD

$$= \sqrt{\text{'Due to testing' mean square}}$$

$$= \sqrt{60.33}$$

$$= 7.77$$

Between-days SD

$$= \sqrt{\frac{\text{Between-days mean square} - \text{Within-days mean square}}{\text{Number of tests per day}}}$$

$$= \sqrt{\frac{701.45 - 60.33}{3.8}}$$

$$= 13.73$$

(c) The sample-to-sample SD is the greater. Sample-to-sample variability will be expected to contribute more than the test error to the overall variability.

(d) SD (for the precision of the mean):

Strategy (i): $\sqrt{\dfrac{13.73^2}{2} + \dfrac{7.77^2}{2 \times 13}} = 9.83$

Strategy (ii): $\sqrt{\dfrac{13.73^2}{5} + \dfrac{7.77^2}{5 \times 2}} = 6.61$

Strategy (iii): $\sqrt{\dfrac{13.73^2}{1} + \dfrac{7.77^2}{1 \times 6}} = 6.44$

The best strategy is the one with the smallest SD.

Although strategy (iii) is marginally better, strategy (ii) has the advantage of making duplicate determinations and therefore of safeguarding against outliers.

Here is the *Crunch* output for Problem 15.2.

Significance Level	5.0%

One - Way Analysis of Variance Table							
Source	Sum of Squares	df	Mean Square	SD	F	Table Value	P
Between Samples	2805.76	4	701.44	13.73	11.63	3.26	0.000
Within Samples	724	12	60.33	7.77			
Total	3529.76	16		15.78			

Adjust Rounding	▲▼		▲▼	▲▼			

Adjust Rounding	▲▼						
Mean	115	91	96	113	127		
Standard Deviation	4.36	4.24	9.17	7.62	9.03		
Observations	3	2	3	4	5		

Data Entry (upto 40 batches) >>>							
Sample							
	A	B	C	D	E		
Test	120	88	104	106	126		
	112	94	98	118	119		
	113		86	107	139		
				121	118		
					133		

Precision of Mean Calculator

No. of Samples	2
No.of Tests per Sample	13
SD (Precision of Mean)	9.83

Chapter 16

16.1 (a) **Analysing percentage yield**

Possible change points occur at observations 12, 16, 26 and 49.

Observations	Test value	Table value	Obs. no. at max. cusum	
1–12	1.2	4.3 (5%)		Not significant
1–16	3.3	5.0 (5%)		Not significant
1–26	6.7	6.2 (5%)	16	Significant
17–26	2.4	3.9 (5%)		Not significant
17–49	10.3	8.4 (1%)	26	Significant
27–49	2.1	5.9 (5%)		Not significant
27–60	11.1	8.5 (1%)	49	Significant
50–60	3.4	4.1 (5%)		Not significant

Stage	Mean yield
1–16	94.0
17–26	92.2
27–49	94.7
50–60	92.0

Analysing catalyst quantity
Possible change points occur at observations 13, 25 and 47.

Observations	Test value	Table value	Obs. no. at max. cusum	
1–13	2.2	4.5 (5%)		Not significant
1–25	14.9	7.3 (1%)	13	Significant
14–25	1.1	4.3 (5%)		Not significant
14–47	20.5	8.5 (1%)	25	Significant
26–47	2.0	5.8 (5%)		Not significant
26–60	16.7	8.6 (1%)	47	Significant
48–60	2.7	4.5 (5%)		Not significant

Stage Stage	Mean catalyst quantity
1–13	2414
14–25	2343
26–47	2425
48–60	2360

Analysing catalyst activity
Possible change points occur at observations 24 and 47.

Observations	Test value	Table value	Obs. no. at max. cusum	
1–24	2.2	6.0 (5%)		Not significant
1–39	11.2	9.2 (1%)	24	Significant
25–39	3.4	4.8 (5%)		Not significant
25–60	13.8	8.8 (1%)	47	Significant
25–47	4.3	5.9 (5%)		Not significant
48–60	2.5	4.5 (5%)		Not significant

Stage	Mean catalyst activity
1–24	135.4
25–47	139.3
48–60	143.4

Appendix

Statistical tables

Statistical Methods in Practice: for Scientists and Technologists   R. Boddy, G. Smith,
© 2009 John Wiley & Sons, Ltd

Table A.1   Normal distribution probabilities

Normal distribution
Probabilities (two-tailed)

z	Prob.	z	Prob.	z	Prob.	z	Prob.	z	Prob.	z	Prob.	z	Prob.	z	Prob.
0.00	1.0000	0.31	0.7566	0.62	0.5353	0.93	0.3524	1.24	0.2150	1.55	0.1211	1.86	0.0629	2.85	0.00437
0.01	0.9920	0.32	0.7490	0.63	0.5287	0.94	0.3472	1.25	0.2113	1.56	0.1188	1.87	0.0615	2.90	0.00373
0.02	0.9840	0.33	0.7414	0.64	0.5222	0.95	0.3421	1.26	0.2077	1.57	0.1164	1.88	0.0601	2.95	0.00318
0.03	0.9761	0.34	0.7339	0.65	0.5157	0.96	0.3371	1.27	0.2041	1.58	0.1141	1.89	0.0588	3.00	0.00270
0.04	0.9681	0.35	0.7263	0.66	0.5093	0.97	0.3320	1.28	0.2005	1.59	0.1118	1.90	0.0574	3.05	0.00229
0.05	0.9601	0.36	0.7188	0.67	0.5029	0.98	0.3271	1.29	0.1971	1.60	0.1096	1.91	0.0561	3.10	0.00194
0.06	0.9522	0.37	0.7114	0.68	0.4965	0.99	0.3222	1.30	0.1936	1.61	0.1074	1.92	0.0549	3.15	0.00163
0.07	0.9442	0.38	0.7039	0.69	0.4902	1.00	0.3173	1.31	0.1902	1.62	0.1052	1.93	0.0536	3.20	0.00137
0.08	0.9362	0.39	0.6965	0.70	0.4839	1.01	0.3125	1.32	0.1868	1.63	0.1031	1.94	0.0524	3.25	0.00115
0.09	0.9283	0.40	0.6892	0.71	0.4777	1.02	0.3077	1.33	0.1835	1.64	0.1010	1.95	0.0512	3.30	0.00097
0.10	0.9203	0.41	0.6818	0.72	0.4715	1.03	0.3030	1.34	0.1802	1.65	0.0989	1.96	0.0500	3.35	0.00081
0.11	0.9124	0.42	0.6745	0.73	0.4654	1.04	0.2983	1.35	0.1770	1.66	0.0969	1.97	0.0488	3.40	0.00067
0.12	0.9045	0.43	0.6672	0.74	0.4593	1.05	0.2937	1.36	0.1738	1.67	0.0949	1.98	0.0477	3.45	0.00056
0.13	0.8966	0.44	0.6599	0.75	0.4533	1.06	0.2891	1.37	0.1707	1.68	0.0930	1.99	0.0466	3.50	0.00047
0.14	0.8887	0.45	0.6527	0.76	0.4473	1.07	0.2846	1.38	0.1676	1.69	0.0910	2.00	0.0455	3.55	0.00039
0.15	0.8808	0.46	0.6455	0.77	0.4413	1.08	0.2801	1.39	0.1645	1.70	0.0891	2.05	0.0404	3.60	0.00032
0.16	0.8729	0.47	0.6384	0.78	0.4354	1.09	0.2757	1.40	0.1615	1.71	0.0873	2.10	0.0357	3.65	0.00026
0.17	0.8650	0.48	0.6312	0.79	0.4295	1.10	0.2713	1.41	0.1585	1.72	0.0854	2.15	0.0316	3.70	0.00022
0.18	0.8572	0.49	0.6241	0.80	0.4237	1.11	0.2670	1.42	0.1556	1.73	0.0836	2.20	0.0278	3.75	0.00018
0.19	0.8493	0.50	0.6171	0.81	0.4179	1.12	0.2627	1.43	0.1527	1.74	0.0819	2.25	0.0244	3.80	0.00014
0.20	0.8415	0.51	0.6101	0.82	0.4122	1.13	0.2585	1.44	0.1499	1.75	0.0801	2.30	0.0214	3.85	0.00012
0.21	0.8337	0.52	0.6031	0.83	0.4065	1.14	0.2543	1.45	0.1471	1.76	0.0784	2.35	0.0188	3.90	0.00010
0.22	0.8259	0.53	0.5961	0.84	0.4009	1.15	0.2501	1.46	0.1443	1.77	0.0767	2.40	0.0164	3.95	0.00008
0.23	0.8181	0.54	0.5892	0.85	0.3953	1.16	0.2460	1.47	0.1416	1.78	0.0751	2.45	0.0143	4.00	6.3×10⁻⁵
0.24	0.8103	0.55	0.5823	0.86	0.3898	1.17	0.2420	1.48	0.1389	1.79	0.0735	2.50	0.0124	4.50	6.8×10⁻⁶
0.25	0.8026	0.56	0.5755	0.87	0.3843	1.18	0.2380	1.49	0.1362	1.80	0.0719	2.55	0.0108	5.00	5.7×10⁻⁷
0.26	0.7949	0.57	0.5687	0.88	0.3789	1.19	0.2340	1.50	0.1336	1.81	0.0703	2.60	0.0093	5.50	3.8×10⁻⁸
0.27	0.7872	0.58	0.5619	0.89	0.3735	1.20	0.2301	1.51	0.1310	1.82	0.0688	2.65	0.0080	6.00	2.0×10⁻⁹
0.28	0.7795	0.59	0.5552	0.90	0.3681	1.21	0.2263	1.52	0.1285	1.83	0.0672	2.70	0.0069		
0.29	0.7718	0.60	0.5485	0.91	0.3628	1.22	0.2225	1.53	0.1260	1.84	0.0658	2.75	0.0060		
0.30	0.7642	0.61	0.5419	0.92	0.3576	1.23	0.2187	1.54	0.1236	1.85	0.0643	2.80	0.0051		

Percentage points (two-tailed)

Significance Level	10% (0.10)	5% (0.05)	2% (0.02)	1% (0.01)	0.2% (0.002)	0.1% (0.001)
z	1.64	1.96	2.33	2.58	3.09	3.29

$$z = \frac{\text{observed value} - \text{population mean}}{\text{population SD}}$$

Table A.1   Normal distribution probabilities (continued)

Normal distribution
Probabilities (one-tailed)

z	Prob.	z	Prob.	z	Prob.	z	Prob.	z	Prob.	z	Prob.	z	Prob.	z	Prob.
0.00	0.5000	0.31	0.3783	0.62	0.2676	0.93	0.1762	1.24	0.1075	1.55	0.0606	1.86	0.0314	2.85	0.00219
0.01	0.4960	0.32	0.3745	0.63	0.2643	0.94	0.1736	1.25	0.1056	1.56	0.0594	1.87	0.0307	2.90	0.00187
0.02	0.4920	0.33	0.3707	0.64	0.2611	0.95	0.1711	1.26	0.1038	1.57	0.0582	1.88	0.0301	2.95	0.00159
0.03	0.4880	0.34	0.3669	0.65	0.2578	0.96	0.1685	1.27	0.1020	1.58	0.0571	1.89	0.0294	3.00	0.00135
0.04	0.4840	0.35	0.3632	0.66	0.2546	0.97	0.1660	1.28	0.1003	1.59	0.0559	1.90	0.0287	3.05	0.00114
0.05	0.4801	0.36	0.3594	0.67	0.2514	0.98	0.1635	1.29	0.0985	1.60	0.0548	1.91	0.0281	3.10	0.00097
0.06	0.4761	0.37	0.3557	0.68	0.2483	0.99	0.1611	1.30	0.0968	1.61	0.0537	1.92	0.0274	3.15	0.00082
0.07	0.4721	0.38	0.3520	0.69	0.2451	1.00	0.1587	1.31	0.0951	1.62	0.0526	1.93	0.0268	3.20	0.00069
0.08	0.4681	0.39	0.3483	0.70	0.2420	1.01	0.1562	1.32	0.0934	1.63	0.0516	1.94	0.0262	3.25	0.00058
0.09	0.4641	0.40	0.3446	0.71	0.2389	1.02	0.1539	1.33	0.0918	1.64	0.0505	1.95	0.0256	3.30	0.00048
0.10	0.4602	0.41	0.3409	0.72	0.2358	1.03	0.1515	1.34	0.0901	1.65	0.0495	1.96	0.0250	3.35	0.00040
0.11	0.4562	0.42	0.3372	0.73	0.2327	1.04	0.1492	1.35	0.0885	1.66	0.0485	1.97	0.0244	3.40	0.00034
0.12	0.4522	0.43	0.3336	0.74	0.2296	1.05	0.1469	1.36	0.0869	1.67	0.0475	1.98	0.0239	3.45	0.00028
0.13	0.4483	0.44	0.3300	0.75	0.2266	1.06	0.1446	1.37	0.0853	1.68	0.0465	1.99	0.0233	3.50	0.00023
0.14	0.4443	0.45	0.3264	0.76	0.2236	1.07	0.1423	1.38	0.0838	1.69	0.0455	2.00	0.0228	3.55	0.00019
0.15	0.4404	0.46	0.3228	0.77	0.2206	1.08	0.1401	1.39	0.0823	1.70	0.0446	2.05	0.0202	3.60	0.00016
0.16	0.4364	0.47	0.3192	0.78	0.2177	1.09	0.1379	1.40	0.0808	1.71	0.0436	2.10	0.0179	3.65	0.00013
0.17	0.4325	0.48	0.3156	0.79	0.2148	1.10	0.1357	1.41	0.0793	1.72	0.0427	2.15	0.0158	3.70	0.00011
0.18	0.4286	0.49	0.3121	0.80	0.2119	1.11	0.1335	1.42	0.0778	1.73	0.0418	2.20	0.0139	3.75	0.00009
0.19	0.4247	0.50	0.3085	0.81	0.2090	1.12	0.1314	1.43	0.0764	1.74	0.0409	2.25	0.0122	3.80	0.00007
0.20	0.4207	0.51	0.3050	0.82	0.2061	1.13	0.1292	1.44	0.0749	1.75	0.0401	2.30	0.0107	3.85	0.00006
0.21	0.4168	0.52	0.3015	0.83	0.2033	1.14	0.1271	1.45	0.0735	1.76	0.0392	2.35	0.0094	3.90	0.00005
0.22	0.4129	0.53	0.2981	0.84	0.2005	1.15	0.1251	1.46	0.0721	1.77	0.0384	2.40	0.0082	3.95	0.00004
0.23	0.4090	0.54	0.2946	0.85	0.1977	1.16	0.1230	1.47	0.0708	1.78	0.0375	2.45	0.0071	4.00	3.2×10^{-5}
0.24	0.4052	0.55	0.2912	0.86	0.1949	1.17	0.1210	1.48	0.0694	1.79	0.0367	2.50	0.0062	4.50	3.4×10^{-6}
0.25	0.4013	0.56	0.2877	0.87	0.1922	1.18	0.1190	1.49	0.0681	1.80	0.0359	2.55	0.0054	5.00	2.9×10^{-7}
0.26	0.3974	0.57	0.2843	0.88	0.1894	1.19	0.1170	1.50	0.0668	1.81	0.0351	2.60	0.0047	5.50	1.9×10^{-8}
0.27	0.3936	0.58	0.2810	0.89	0.1867	1.20	0.1151	1.51	0.0655	1.82	0.0344	2.65	0.0040	6.00	1.0×10^{-9}
0.28	0.3897	0.59	0.2776	0.90	0.1841	1.21	0.1131	1.52	0.0643	1.83	0.0336	2.70	0.0035		
0.29	0.3859	0.60	0.2743	0.91	0.1814	1.22	0.1112	1.53	0.0630	1.84	0.0329	2.75	0.0030		
0.30	0.3821	0.61	0.2709	0.92	0.1788	1.23	0.1093	1.54	0.0618	1.85	0.0322	2.80	0.0026		

Percentage points (one-tailed)

Significance Level	10% (0.10)	5% (0.05)	2½% (0.025)	1% (0.01)	0.5% (0.005)	0.1% (0.001)
z	1.28	1.64	1.96	2.33	2.58	3.09

$$z = \frac{\text{observed value} - \text{population mean}}{\text{population SD}}$$

Table A.2 t-test

Significance level Degrees of freedom	10% (0.1)	5% (0.05)	2% (0.02)	1% (0.01)	0.2% (0.002)	0.1% (0.001)
1	6.31	12.71	31.82	63.66	318.3	636.6
2	2.92	4.30	6.96	9.92	22.33	31.60
3	2.35	3.18	4.54	5.84	10.21	12.92
4	2.13	2.78	3.75	4.60	7.17	8.61
5	2.02	2.57	3.36	4.03	5.89	6.87
6	1.94	2.45	3.14	3.71	5.21	5.96
7	1.89	2.36	3.00	3.50	4.79	5.41
8	1.86	2.31	2.90	3.36	4.50	5.04
9	1.83	2.26	2.82	3.25	4.30	4.78
10	1.81	2.23	2.76	3.17	4.14	4.59
11	1.80	2.20	2.72	3.11	4.02	4.44
12	1.78	2.18	2.68	3.05	3.93	4.32
13	1.77	2.16	2.65	3.01	3.85	4.22
14	1.76	2.14	2.62	2.98	3.79	4.14
15	1.75	2.13	2.60	2.95	3.73	4.07
16	1.75	2.12	2.58	2.92	3.69	4.01
17	1.74	2.11	2.57	2.90	3.65	3.97
18	1.73	2.10	2.55	2.88	3.61	3.92
19	1.73	2.09	2.54	2.86	3.58	3.88
20	1.72	2.09	2.53	2.85	3.55	3.85
25	1.71	2.06	2.49	2.79	3.45	3.73
30	1.70	2.04	2.46	2.75	3.39	3.65
40	1.68	2.02	2.42	2.70	3.31	3.55
60	1.67	2.00	2.39	2.66	3.23	3.46
120	1.66	1.98	2.36	2.62	3.16	3.37
Infinity	1.64	1.96	2.33	2.58	3.09	3.29
Confidence level	90%	95%	98%	99%	99.8	99.9%

Note: The quoted significance levels are for a two-sided test. To carry out a one-sided test, halve the significance level given at the top of the table.

Table A.3   Grubbs' test for a single outlier using mean and standard deviation

Significance level Degrees of freedom	5% (0.05)	1% (0.01)
2	1.15	1.15
3	1.48	1.50
4	1.71	1.76
5	1.89	1.97
6	2.02	2.14
7	2.13	2.27
8	2.21	2.39
9	2.29	2.48
10	2.36	2.56
11	2.41	2.64
12	2.46	2.70
13	2.51	2.76
14	2.55	2.81
15	2.59	2.85
16	2.62	2.89
17	2.65	2.93
18	2.68	2.97
19	2.71	3.00
20	2.73	3.03
30	2.92	3.26
40	3.05	3.39
50	3.14	3.49
100	3.38	3.75

$$\text{Test value} = \frac{|x - \bar{x}|}{s}$$

where x is the most extreme observation from the mean;

$\bar{x}$ is the mean of all observations including the possible outlier;

s is the standard deviation of all observations including the possible outlier.

Table A.4   Tolerance interval coefficients

Confidence level	90%			95%		
% of items within tolerance interval Sample size	90%	95%	99%	90%	95%	99%
3	5.85	6.92	8.97	8.38	9.92	12.86
4	4.17	4.94	6.44	5.37	6.37	8.30
5	3.49	4.15	5.42	4.28	5.08	6.63
6	3.13	3.72	4.87	3.71	4.41	5.78
7	2.90	3.45	4.52	3.31	4.01	5.25
8	2.74	3.26	4.28	3.14	3.73	4.89
9	2.63	3.13	4.10	2.97	3.53	4.63
10	2.54	3.02	3.96	2.84	3.38	4.43
12	2.40	2.86	3.76	2.66	3.16	4.15
14	2.31	2.76	3.62	2.53	3.01	3.96
16	2.25	2.68	3.51	2.44	2.90	3.81
18	2.19	2.61	3.43	2.37	2.82	3.70
20	2.15	2.56	3.37	2.31	2.75	3.62
30	2.03	2.41	3.17	2.14	2.55	3.35
40	1.96	2.33	3.07	2.05	2.45	3.21
50	1.92	2.28	3.00	2.00	2.38	3.13
Infinity	1.65	1.96	2.58	1.65	1.96	2.58

Tolerance interval $= \bar{x} \pm ks$

Table A.5   Percentages for probability plots (median ranks)

Sample size Rank	1	2	3	4	5	6	7	8	9	10
1	50.0	29.3	20.6	15.9	12.9	10.9	9.4	8.3	7.4	6.7
2		70.7	50.0	38.6	31.5	26.6	23.0	20.2	18.1	16.3
3			79.4	61.4	50.0	42.2	36.5	32.1	28.7	25.9
4				84.1	68.5	57.8	50.0	44.0	39.4	35.6
5					87.1	73.4	63.5	56.0	50.0	45.2
6						89.1	77.0	67.9	60.6	54.8
7							90.6	79.8	71.3	64.4
8								91.7	81.9	74.1
9									92.6	83.7
10										93.3

Sample size Rank	11	12	13	14	15	16	17	18	19	20
1	6.1	5.6	5.2	4.8	4.5	4.2	4.0	3.8	3.6	3.4
2	14.9	13.7	12.7	11.8	11.0	10.3	9.8	9.2	8.7	8.3
3	23.7	21.8	20.1	18.7	17.5	16.4	15.5	14.6	13.9	13.2
4	32.4	29.8	27.6	25.7	24.0	22.5	21.2	20.1	19.0	18.1
5	41.2	37.9	35.1	32.6	30.5	28.6	27.0	25.5	24.2	23.0
6	50.0	46.0	42.5	39.6	37.0	34.8	32.8	31.0	29.4	27.9
7	58.8	54.0	50.0	46.5	43.5	40.8	38.5	36.4	34.5	32.8
8	67.6	62.1	57.5	53.5	50.0	47.0	44.2	41.8	39.7	37.7
9	76.3	70.2	64.9	60.4	56.5	53.0	50.0	47.3	44.8	42.6
10	85.1	78.2	72.4	67.4	63.0	59.2	55.8	52.7	50.0	47.6
11	93.9	86.3	79.9	74.3	69.5	65.2	61.5	58.2	55.2	52.4
12		94.4	87.3	81.3	76.0	71.4	67.2	63.6	60.3	57.4
13			94.8	88.2	82.5	77.5	73.0	69.0	65.5	62.3
14				95.2	89.0	83.6	78.8	74.5	70.6	67.2
15					95.5	89.7	84.5	79.9	75.8	72.1
16						95.8	90.2	85.4	81.0	77.0
17							96.0	90.8	86.1	81.9
18								96.2	91.3	86.8
19									96.4	91.7
20										96.6

For sample sizes above 20:

$$\text{Estimated percentage in population (median rank)} = \frac{100\,(\text{Rank} - 0.3)}{n + 0.4}$$

Table A.6 F-test

F-test (two-sided)

5% (0.05) significance level

Degrees of freedom for smaller SD	Degrees of freedom for larger SD														
	1	2	3	4	5	6	7	8	9	10	12	15	20	60	Infinity
1	647.8	799.5	864.2	899.6	921.8	937.1	948.2	956.7	963.3	968.6	976.7	984.9	993.1	1010	1018
2	38.51	39.00	39.17	39.25	39.30	39.33	39.36	39.37	39.39	39.40	39.41	39.43	39.45	39.48	39.50
3	17.44	16.04	15.44	15.10	14.88	14.73	14.62	14.54	14.47	14.42	14.34	14.25	14.17	13.99	13.90
4	12.22	10.65	9.98	9.60	9.36	9.20	9.07	8.98	8.90	8.84	8.75	8.66	8.56	8.36	8.26
5	10.01	8.43	7.76	7.39	7.15	6.98	6.85	6.76	6.68	6.62	6.52	6.43	6.33	6.12	6.02
6	8.81	7.26	6.60	6.23	5.99	5.82	5.70	5.60	5.52	5.46	5.37	5.27	5.17	4.96	4.85
7	8.07	6.54	5.89	5.52	5.29	5.12	4.99	4.90	4.82	4.76	4.67	4.57	4.47	4.25	4.14
8	7.57	6.06	5.42	5.05	4.82	4.65	4.53	4.43	4.36	4.30	4.20	4.10	4.00	3.78	3.67
9	7.21	5.71	5.08	4.72	4.48	4.32	4.20	4.10	4.03	3.96	3.87	3.77	3.67	3.45	3.33
10	6.94	5.46	4.83	4.47	4.24	4.07	3.95	3.85	3.78	3.72	3.62	3.52	3.42	3.20	3.08
12	6.55	5.10	4.47	4.12	3.89	3.73	3.61	3.51	3.44	3.37	3.28	3.18	3.07	2.85	2.72
15	6.20	4.77	4.15	3.80	3.58	3.41	3.29	3.20	3.12	3.06	2.96	2.86	2.76	2.52	2.40
20	5.87	4.46	3.86	3.51	3.29	3.13	3.01	2.91	2.84	2.77	2.68	2.57	2.46	2.22	2.09
60	5.29	3.93	3.34	3.01	2.79	2.63	2.51	2.41	2.33	2.27	2.17	2.06	1.94	1.67	1.48
Infinity	5.02	3.69	3.12	2.79	2.57	2.41	2.29	2.19	2.11	2.05	1.94	1.83	1.71	1.39	1.00

1% (0.01) significance level

Degrees of freedom for smaller SD	Degrees of freedom for larger SD														
	1	2	3	4	5	6	7	8	9	10	12	15	20	60	Infinity
1	16211	20000	21615	22500	23056	23437	23715	23925	24091	24224	24426	24630	24836	25253	25465
2	198.5	199.0	199.2	199.2	199.3	199.3	199.4	199.4	199.4	199.4	199.4	199.4	199.4	199.5	199.5
3	55.55	49.80	47.47	46.19	45.39	44.84	44.43	44.13	43.88	43.69	43.29	43.08	42.78	42.15	41.83
4	31.33	26.28	24.26	23.15	22.46	21.97	21.62	21.35	21.14	20.97	20.70	20.04	20.17	19.61	19.32
5	22.78	18.31	16.53	15.56	14.94	14.51	14.20	13.96	13.77	13.62	13.38	13.15	12.90	12.40	12.14
6	18.63	14.54	12.92	12.03	11.46	11.07	10.79	10.57	10.39	10.25	10.03	9.81	9.59	9.12	8.88
7	16.24	12.40	10.88	10.05	9.52	9.16	8.89	8.68	8.51	8.38	8.18	7.97	7.75	7.31	7.08
8	14.69	11.04	9.60	8.81	8.30	7.95	7.69	7.50	7.34	7.21	7.01	6.81	6.61	6.18	5.95
9	13.61	10.11	8.72	7.96	7.47	7.13	6.88	6.69	6.54	6.42	6.23	6.03	5.83	5.41	5.19
10	12.83	9.43	8.08	7.34	6.87	6.54	6.30	6.12	5.97	5.85	5.66	5.47	5.27	4.86	4.64
12	11.75	8.51	7.23	6.52	6.07	5.76	5.52	5.35	5.20	5.09	4.91	4.72	4.53	4.12	3.90
15	10.80	7.70	6.48	5.80	5.37	5.07	4.85	4.67	4.54	4.42	4.25	4.07	3.88	3.48	3.26
20	9.94	6.99	5.82	5.17	4.76	4.47	4.26	4.09	3.96	3.85	3.68	3.50	3.32	2.92	2.69
60	8.49	5.79	4.73	4.14	3.76	3.49	3.29	3.13	3.01	2.90	2.74	2.57	2.39	1.96	1.69
Infinity	7.88	5.30	4.28	3.72	3.35	3.09	2.90	2.74	2.62	2.52	2.36	2.19	2.00	1.53	1.00

$$\text{Test value} = \frac{\text{Larger } S^2}{\text{Smaller } S^2}$$

Table A.6 F-test (continued)

F-test (one-sided)
for use in analysis of variance

5% (0.05) significance level

Degrees of freedom for residual mean square	Degrees of freedom for treatment mean square														
	1	2	3	4	5	6	7	8	9	10	12	15	20	60	Infinity
1	161.4	199.5	215.7	224.6	230.2	234.0	236.8	238.9	240.5	241.9	243.9	246.0	248.0	252.2	254.3
2	18.51	19.00	19.16	19.25	19.30	19.33	19.35	19.37	19.38	19.40	19.41	19.43	19.45	19.48	19.50
3	10.13	9.55	9.28	9.12	9.01	8.94	8.89	8.85	8.81	8.79	8.74	8.70	8.66	8.57	8.53
4	7.71	6.94	6.59	6.39	6.26	6.16	6.09	6.04	6.00	5.96	5.91	5.86	5.80	5.69	5.63
5	6.61	5.79	5.41	5.19	5.05	4.95	4.88	4.82	4.77	4.74	4.68	4.62	4.56	4.43	4.36
6	5.99	5.14	4.76	4.53	4.39	4.28	4.21	4.15	4.10	4.06	4.00	3.94	3.87	3.74	3.67
7	5.59	4.74	4.35	4.12	3.97	3.87	3.79	3.73	3.68	3.64	3.57	3.51	3.44	3.30	3.23
8	5.32	4.46	4.07	3.84	3.69	3.58	3.50	3.44	3.39	3.35	3.28	3.22	3.15	3.01	2.93
9	5.12	4.26	3.86	3.63	3.48	3.37	3.29	3.23	3.18	3.14	3.07	3.01	2.94	2.79	2.71
10	4.96	4.10	3.71	3.48	3.33	3.22	3.14	3.07	3.02	2.98	2.91	2.85	2.77	2.62	2.54
12	4.75	3.89	3.49	3.26	3.11	3.00	2.91	2.85	2.80	2.75	2.69	2.62	2.54	2.38	2.30
15	4.54	3.68	3.29	3.06	2.90	2.79	2.71	2.64	2.59	2.54	2.48	2.40	2.33	2.16	2.07
20	4.35	3.49	3.10	2.87	2.71	2.60	2.49	2.45	2.39	2.35	2.28	2.20	2.12	1.95	1.84
60	4.00	3.15	2.76	2.53	2.37	2.25	2.17	2.10	2.04	1.99	1.92	1.84	1.75	1.53	1.39
Infinity	3.84	3.00	2.60	2.37	2.21	2.10	2.01	1.94	1.88	1.83	1.75	1.67	1.57	1.32	1.00

1% (0.01) significance level

Degrees of freedom for residual mean square	Degrees of freedom for treatment mean square														
	1	2	3	4	5	6	7	8	9	10	12	15	20	60	Infinity
1	4052	5000	5403	5625	5764	5859	5928	5982	6022	6056	6106	6157	6209	6313	6366
2	98.50	99.00	99.17	99.25	99.30	99.33	99.36	99.37	99.39	99.40	99.42	99.43	99.45	99.48	99.50
3	34.12	30.82	29.46	28.71	28.24	27.91	27.67	27.49	27.35	27.23	27.05	26.87	26.69	26.32	26.13
4	21.20	18.00	16.69	15.98	15.52	15.21	14.98	14.80	14.66	14.55	14.37	14.20	14.02	13.65	13.46
5	16.26	13.27	12.06	11.39	10.97	10.67	10.46	10.29	10.16	10.05	9.89	9.72	9.55	9.20	9.02
6	13.75	10.92	9.78	9.15	8.75	8.47	8.26	8.10	7.98	7.87	7.72	7.56	7.40	7.06	6.88
7	12.25	9.55	8.45	7.85	7.46	7.19	6.99	6.84	6.72	6.62	6.47	6.31	6.16	5.82	5.65
8	11.26	8.65	7.59	7.01	6.63	6.37	6.18	6.03	5.91	5.81	5.67	5.52	5.36	5.03	4.86
9	10.56	8.02	6.99	6.42	6.06	5.80	5.61	5.47	5.35	5.26	5.11	4.96	4.81	4.48	4.31
10	10.04	7.56	6.55	5.99	5.64	5.39	5.20	5.06	4.94	4.85	4.71	4.56	4.41	4.08	3.91
12	9.33	6.93	5.95	5.41	5.06	4.82	4.64	4.50	4.39	4.30	4.16	4.01	3.86	3.54	3.36
15	8.68	6.36	5.42	4.89	4.56	4.32	4.14	4.00	3.89	3.80	3.67	3.52	3.37	3.05	2.87
20	8.10	5.85	4.94	4.43	4.10	3.87	3.70	3.56	3.46	3.37	3.23	3.09	2.94	2.61	2.42
60	7.08	4.98	4.13	3.65	3.34	3.12	2.95	2.82	2.72	2.63	2.50	2.35	2.20	1.84	1.60
Infinity	6.63	4.61	3.78	3.32	3.02	2.80	2.64	2.51	2.41	2.32	2.18	2.04	1.88	1.47	1.00

$$\text{Test value} = \frac{\text{Effect MS}}{\text{Residual MS}}$$

Table A.7   Confidence interval for a population standard deviation

Level of Confidence Degrees of freedom	90%			95%			99%		
	k_a	k_b	k_b/k_a	k_a	k_b	k_b/k_a	k_a	k_b	k_b/k_a
1	0.51	15.9	31.18	0.45	31.9	70.89	0.36	160	444.44
2	0.58	4.41	7.60	0.52	6.28	12.08	0.43	14.1	32.79
3	0.62	2.92	4.71	0.57	3.73	6.54	0.48	6.47	13.48
4	0.65	2.37	3.65	0.60	2.87	4.78	0.52	4.40	8.46
5	0.67	2.09	3.12	0.62	2.45	3.95	0.55	3.48	6.33
6	0.69	1.92	2.78	0.64	2.20	3.44	0.57	2.98	5.23
7	0.71	1.80	2.54	0.66	2.04	3.09	0.59	2.66	4.51
8	0.72	1.71	2.38	0.68	1.92	2.82	0.60	2.44	4.07
9	0.73	1.65	2.26	0.69	1.83	2.65	0.62	2.28	3.68
10	0.74	1.59	2.15	0.70	1.75	2.50	0.63	2.15	3.41
12	0.76	1.52	2.00	0.72	1.65	2.29	0.65	1.98	3.05
15	0.77	1.44	1.87	0.74	1.55	2.09	0.68	1.81	2.66
20	0.80	1.36	1.70	0.77	1.44	1.87	0.71	1.64	2.31
24	0.81	1.32	1.63	0.78	1.39	1.78	0.73	1.56	2.14
30	0.83	1.27	1.53	0.80	1.34	1.68	0.75	1.48	1.97
40	0.85	1.23	1.45	0.82	1.28	1.56	0.77	1.39	1.81
60	0.87	1.18	1.36	0.85	1.22	1.44	0.81	1.30	1.60
Infinity	1.00	1.00	1.00	1.00	1.00	1.00	1.00	1.00	1.00

Confidence interval is $k_a s$ to $k_b s$

Table A.8   Chi-squared tests

	Chi-squared values for contingency tables and goodness of fit		Modified chi-squared values for Poisson dispersion test					
	One-sided		One-sided		Two-sided			
Significance level					5% (0.05)		1% (0.01)	
Degrees of freedom	5% (0.05)	1% (0.01)	5% (0.05)	1% (0.01)	Lower value	Upper value	Lower value	Upper value
1	3.84	6.63	3.84	6.63	0.00	5.02	0.00	7.88
2	5.99	9.21	3.00	4.60	0.02	3.69	0.00	5.30
3	7.81	11.34	2.61	3.78	0.07	3.12	0.02	4.28
4	9.49	13.28	2.37	3.32	0.12	2.79	0.05	3.72
5	11.07	15.08	2.21	3.02	0.17	2.57	0.08	3.35
6	12.59	16.81	2.10	2.80	0.21	2.41	0.11	3.09
7	14.07	18.49	2.01	2.64	0.24	2.29	0.14	2.90
8	15.51	20.09	1.94	2.51	0.27	2.19	0.17	2.74
9	16.92	21.67	1.88	2.41	0.30	2.11	0.19	2.62
10	18.31	23.21	1.83	2.32	0.32	2.05	0.22	2.52
11	19.68	24.72	1.79	2.25	0.35	1.99	0.24	2.43
12	21.03	26.22	1.75	2.18	0.37	1.94	0.26	2.36
13	22.36	27.69	1.72	2.13	0.39	1.90	0.27	2.29
14	23.68	29.14	1.69	2.08	0.40	1.87	0.29	2.24
15	25.00	30.58	1.67	2.04	0.42	1.83	0.31	2.19
16	26.30	32.00	1.64	2.00	0.43	1.80	0.32	2.14
17	27.59	33.41	1.62	1.97	0.44	1.78	0.34	2.10
18	28.87	34.81	1.60	1.93	0.46	1.75	0.35	2.06
19	30.14	36.19	1.59	1.90	0.47	1.73	0.36	2.03
20	31.41	37.57	1.57	1.88	0.48	1.71	0.37	2.00
25	37.65	44.31	1.51	1.77	0.52	1.63	0.42	1.88
30	43.77	50.89	1.46	1.70	0.56	1.57	0.46	1.79
40	55.76	63.69	1.39	1.59	0.61	1.48	0.52	1.67
50	67.50	76.15	1.35	1.52	0.65	1.43	0.56	1.59
60	79.08	88.38	1.32	1.47	0.67	1.39	0.59	1.53
70	90.53	100.42	1.29	1.43	0.70	1.36	0.62	1.49
80	101.88	112.33	1.27	1.40	0.71	1.33	0.64	1.45
90	113.14	124.12	1.26	1.38	0.73	1.31	0.66	1.43
100	124.34	135.81	1.24	1.36	0.74	1.30	0.67	1.40

Goodness of fit, contingency   Test value $= \sum \dfrac{(O - E)^2}{E}$

Poisson dispersion test   Test value $= \dfrac{SD^2}{Mean}$

Table A.9   Goldsmith's cusum span test

Significance level Length of span	10% (0.10)	5% (0.05)	1% (0.01)
5	2.4	2.7	3.3
6	2.7	3.0	3.6
7	2.9	3.2	4.0
8	3.2	3.5	4.3
9	3.4	3.7	4.6
10	3.6	3.9	4.9
11	3.8	4.1	5.1
12	3.9	4.3	5.3
13	4.0	4.5	5.5
14	4.1	4.6	5.6
15	4.2	4.8	5.8
20	5.2	5.6	6.8
25	5.6	6.0	7.3
30	6.2	6.7	8.0
40	7.2	7.8	9.3
50	8.0	8.6	10.4
60	8.8	9.5	11.3
70	9.5	10.3	12.2
80	10.1	10.8	12.9
90	10.5	11.3	13.6
100	11.0	11.8	14.3

$$\text{Test value} = \frac{\text{Maximum} |\text{cusum}|}{\text{Localised SD}}$$

Table values obtained from Nomogram in BS 5703 Part II which was devised from values obtained by simulation.

Table A.10 Binomial distribution probabilities

n = 5

Def. in sample	Def. in batch							
	0.1%	0.2%	0.5%	1%	2%	5%	10%	20%
0	0.9950	0.9900	0.975	0.951	0.904	0.774	0.591	0.328
1	0.0050	0.0099	0.025	0.048	0.092	0.204	0.328	0.410
2	0.0000	0.0001	0.000	0.001	0.004	0.021	0.073	0.205
3	0.0000	0.0001	0.000	0.000	0.000	0.001	0.008	0.051

n = 10

Def. in sample	Def. in batch							
	0.1%	0.2%	0.5%	1%	2%	5%	10%	20%
0	0.9900	0.9802	0.951	0.904	0.817	0.599	0.349	0.108
1	0.0099	0.0196	0.048	0.092	0.167	0.315	0.387	0.268
2	0.0001	0.0002	0.001	0.004	0.015	0.074	0.194	0.302
3	0.0000	0.0000	0.000	0.000	0.001	0.011	0.057	0.201

n = 20

Def. in sample	Def. in batch							
	0.1%	0.2%	0.5%	1%	2%	5%	10%	20%
0	0.9802	0.9608	0.905	0.818	0.668	0.359	0.122	0.012
1	0.0196	0.0385	0.091	0.165	0.272	0.377	0.270	0.058
2	0.0002	0.0007	0.004	0.016	0.053	0.189	0.285	0.137
3	0.0000	0.0000	0.000	0.001	0.006	0.060	0.190	0.205

n = 30

Def. in sample	Def. in batch							
	0.1%	0.2%	0.5%	1%	2%	5%	10%	20%
0	0.9704	0.9417	0.860	0.740	0.545	0.215	0.042	0.001
1	0.0291	0.0566	0.130	0.224	0.334	0.339	0.141	0.009
2	0.0005	0.0016	0.009	0.033	0.099	0.258	0.228	0.034
3	0.0000	0.0001	0.001	0.003	0.019	0.127	0.236	0.079
4	0.0000	0.0000	0.000	0.000	0.003	0.045	0.177	0.132

n = 40

Def. in sample	Def. in batch							
	0.1%	0.2%	0.5%	1%	2%	5%	10%	20%
0	0.961	0.923	0.818	0.669	0.446	0.129	0.015	0.000
1	0.038	0.074	0.164	0.270	0.364	0.271	0.066	0.001
2	0.001	0.003	0.016	0.053	0.145	0.278	0.142	0.006
3	0.000	0.000	0.002	0.007	0.037	0.185	0.200	0.021
4	0.000	0.000	0.000	0.001	0.007	0.090	0.206	0.047

n = 50

Def. in sample	Def. in batch							
	0.1%	0.2%	0.5%	1%	2%	5%	10%	20%
0	0.951	0.905	0.778	0.605	0.364	0.077	0.005	0.000
1	0.048	0.091	0.196	0.306	0.371	0.202	0.029	0.000
2	0.001	0.004	0.024	0.076	0.186	0.261	0.078	0.001
3	0.000	0.000	0.002	0.012	0.061	0.220	0.139	0.004
4	0.000	0.000	0.000	0.001	0.015	0.136	0.181	0.013

n = 80

Def. in sample	Def. in batch							
	0.1%	0.2%	0.5%	1%	2%	5%	10%	20%
0	0.923	0.852	0.670	0.448	0.199	0.017	0.000	0.000
1	0.074	0.137	0.269	0.362	0.324	0.070	0.002	0.000
2	0.003	0.011	0.053	0.144	0.261	0.144	0.009	0.000
3	0.000	0.000	0.007	0.038	0.139	0.197	0.025	0.000
4	0.000	0.000	0.001	0.007	0.054	0.200	0.053	0.000
5	0.000	0.000	0.000	0.001	0.017	0.160	0.089	0.000

n = 100

Def. in sample	Def. in batch							
	0.1%	0.2%	0.5%	1%	2%	5%	10%	20%
0	0.905	0.819	0.606	0.366	0.133	0.006	0.000	0.000
1	0.091	0.164	0.304	0.370	0.271	0.031	0.000	0.000
2	0.004	0.016	0.076	0.185	0.273	0.081	0.002	0.000
3	0.000	0.001	0.012	0.061	0.182	0.140	0.006	0.000
4	0.000	0.000	0.002	0.015	0.090	0.178	0.016	0.000
5	0.000	0.000	0.000	0.003	0.035	0.180	0.034	0.000

Table A.11   Poisson distribution probabilities

Observed count / Average count	0.1	0.2	0.3	0.4	0.5	0.6	0.7
0	0.9048	0.8187	0.7408	0.6703	0.6065	0.5488	0.4966
1	0.0905	0.1637	0.2222	0.2681	0.3033	0.3293	0.3476
2	0.0045	0.0164	0.0333	0.0536	0.0758	0.0988	0.1217
3	0.0002	0.0011	0.0033	0.0072	0.0126	0.0198	0.0284
4	0.0000	0.0001	0.0002	0.0007	0.0016	0.0030	0.0050

Observed count / Average count	0.8	0.9	1.0	1.2	1.4	1.6	1.8
0	0.4493	0.4066	0.3679	0.3012	0.2466	0.2019	0.1653
1	0.3595	0.3659	0.3679	0.3614	0.3452	0.3230	0.2975
2	0.1438	0.1647	0.1839	0.2169	0.2417	0.2564	0.2678
3	0.0383	0.0494	0.0613	0.0867	0.1128	0.1378	0.1607
4	0.0077	0.0111	0.0153	0.0260	0.0395	0.0551	0.0723
5	0.0012	0.0020	0.0031	0.0062	0.0111	0.0176	0.0260
6	0.0002	0.0003	0.0005	0.0012	0.0026	0.0047	0.0078
7	0.0000	0.0000	0.0001	0.0002	0.0005	0.0011	0.0020

Observed count / Average count	2.0	2.5	3.0	3.5	4.0	4.5	5.0
0	0.1353	0.0821	0.0498	0.0302	0.0183	0.0111	0.0067
1	0.2707	0.2052	0.1494	0.1057	0.0733	0.0500	0.0337
2	0.2707	0.2565	0.2240	0.1850	0.1465	0.1125	0.0842
3	0.1804	0.2138	0.2240	0.2158	0.1954	0.1687	0.1404
4	0.0902	0.1336	0.1680	0.1888	0.1954	0.1898	0.1755
5	0.0361	0.0668	0.1008	0.1322	0.1563	0.1708	0.1755
6	0.0120	0.0278	0.0504	0.0771	0.1042	0.1281	0.1462
7	0.0034	0.0099	0.0216	0.0385	0.0595	0.0824	0.1044
8	0.0009	0.0031	0.0081	0.0169	0.0298	0.0463	0.0653
9	0.0001	0.0009	0.0027	0.0066	0.0132	0.0232	0.0363
10	0.0000	0.0002	0.0008	0.0023	0.0053	0.0104	0.0181

Table A.12   Correlation coefficient test (Pearson's)

Significance level Degrees of freedom	5% (0.05)	1% (0.01)
2	0.950	0.990
3	0.878	0.959
4	0.811	0.917
5	0.754	0.875
6	0.707	0.834
7	0.666	0.798
8	0.632	0.765
9	0.602	0.735
10	0.576	0.708
11	0.553	0.684
12	0.532	0.661
13	0.514	0.641
14	0.497	0.623
15	0.482	0.606
20	0.423	0.537
30	0.349	0.449
40	0.304	0.393
60	0.250	0.325

Test value = |sample correlation coefficient|

Degrees of freedom: 2 less than the sample size

Table A.13   Binomial tests

$\pi = \frac{1}{2}$ (for use with duo-trio, preference, paired comparison and sign test)

Number of trials	Significance level	Two-sided		One-sided	
		5% (0.05)	1% (0.01)	5% (0.05)	1% (0.01)
5		—	—	4½	—
6		5½	—	5½	—
7		6½	—	6½	6½
8		7½	7½	6½	7½
9		7½	8½	7½	8½
10		8½	9½	8½	9½
11		9½	10½	8½	9½
12		9½	10½	9½	10½
13		10½	11½	9½	11½
14		11½	12½	10½	11½
15		11½	12½	11½	12½
16		12½	13½	11½	13½
17		12½	14½	12½	13½
18		13½	14½	12½	14½
19		14½	15½	13½	14½
20		14½	16½	14½	15½
25		17½	19½	17½	18½
30		20½	22½	19½	21½
35		23½	25½	22½	24½
40		26½	28½	25½	27½
45		29½	31½	28½	30½
50		32½	34½	31½	33½
60		38½	40½	36½	39½
70		43½	46½	42½	45½
80		49½	51½	47½	50½
90		54½	57½	53½	56½
100		60½	63½	58½	62½

Table A.13   Binomial tests (continued)

	$\pi = \frac{1}{3}$		$\pi = \frac{1}{10}$	
	(for use with triangle test)		(for use with the two-out-of-five test)	
Significance level	5% (0.05)	1% (0.01)	5% (0.05)	1% (0.01)
Number of trials				
2	—	—	1½	1½
3	2½	—	1½	2½
4	3½	—	2½	2½
5	3½	4½	2½	2½
6	4½	5½	2½	3½
7	4½	5½	2½	3½
8	5½	6½	2½	3½
9	5½	6½	3½	3½
10	6½	7½	3½	4½
11	6½	7½	3½	4½
12	7½	8½	3½	4½
13	7½	8½	3½	4½
14	8½	9½	3½	4½
15	8½	9½	4½	5½
16	8½	10½	4½	5½
17	9½	10½	4½	5½
18	9½	11½	4½	5½
19	10½	11½	4½	5½
20	10½	12½	4½	6½
25	12½	14½	5½	6½
30	14½	16½	6½	7½
35	16½	18½	7½	8½
40	18½	20½	7½	9½
45	20½	23½	8½	10½
50	22½	25½	9½	10½
60	26½	29½	10½	12½
70	30½	33½	11½	13½
80	34½	37½	13½	15½
90	37½	41½	14½	16½
100	41½	45½	15½	17½

Table A.14   Wilcoxon–Mann–Whitney test

Sample sizes		Significance level	
Smaller	Larger	5% (0.05)	1% (0.01)
3	3	—	—
3	4	—	—
3	5	20½	—
3	6	22½	—
3	7	25½	—
3	8	27½	—
3	9	30½	32½
3	10	32½	35½
4	4	25½	—
4	5	28½	—
4	6	31½	33½
4	7	34½	37½
4	8	37½	40½
4	9	41½	44½
4	10	43½	48½
5	5	37½	39½
5	6	41½	43½
5	7	44½	48½
5	8	48½	52½
5	9	52½	56½
5	10	56½	60½
6	6	51½	54½
6	7	56½	59½
6	8	60½	64½
6	9	64½	69½
6	10	69½	74½
7	7	68½	72½
7	8	73½	77½
7	9	78½	83½
7	10	83½	88½
8	8	86½	92½
8	9	92½	98½
8	10	98½	104½
9	9	108½	114½
9	10	114½	121½
10	10	131½	138½

The test value is equal to the rank total of the smaller sample. The procedure for assigning the ranks should be reversed if so doing would increase the rank total of the smaller sample.

Table A.15   Wilcoxon matched-pairs signed ranks test

Significance level Sample size	5% (0.05)	1% (0.01)
5	—	—
6	20½	—
7	25½	—
8	32½	35½
9	39½	43½
10	46½	51½
11	55½	60½
12	64½	70½
13	73½	81½
14	83½	92½
15	94½	104½
16	106½	116½
17	118½	129½
18	130½	143½
19	143½	157½
20	157½	172½

The test value is equal to the larger of the two rank totals.

Table A.16   Confidence limits for a Poisson distribution

Sample count	One-sided		Two-sided			
	95%	99%	95%		99%	
	Lower limit	Upper limit	Lower limit	Upper limit	Lower limit	Upper limit
0	3.0	4.6		3.7		5.3
1	4.7	6.6		5.6		7.4
2	6.3	8.4	0.2	7.2	0.1	9.3
3	7.7	10.0	0.6	8.8	0.3	11.0
4	9.1	11.6	1.1	10.2	0.7	12.6
5	10.5	13.1	1.6	11.7	1.1	14.1
6	11.8	14.6	2.2	13.1	1.5	15.7
7	13.1	16.0	2.8	14.4	2.0	17.1
8	14.4	17.4	3.4	15.8	2.6	18.6
9	15.7	18.8	4.1	17.1	3.1	20.0
10	17.0	20.1	4.8	18.4	3.7	21.4
11	18.2	21.5	5.5	19.7	4.3	22.8
12	19.4	22.8	6.2	21.0	4.9	24.1
13	20.7	24.1	6.9	22.2	5.6	25.5
14	21.9	25.4	7.6	23.5	6.2	26.8
15	23.1	26.7	8.4	24.7	6.9	28.2
16	24.3	28.0	9.1	26.0	7.6	29.5
17	25.5	29.3	9.9	27.2	8.2	30.8
18	26.7	30.6	10.7	28.5	8.9	32.1
19	27.9	31.8	11.4	29.7	9.6	33.4
20	29.1	33.1	12.2	30.9	10.3	34.7
21	30.2	34.4	13.0	32.1	11.1	35.9
22	31.4	35.6	13.8	33.3	11.8	37.2
23	32.6	36.8	14.6	34.5	12.5	38.5
24	33.8	38.1	15.4	35.7	13.3	39.7
25	34.9	39.3	16.2	36.9	14.0	41.0
26	36.1	40.5	17.0	38.1	14.7	42.2
27	37.2	41.8	17.8	39.3	15.5	43.5
28	38.4	43.0	18.6	40.5	16.2	44.7
29	39.5	44.2	19.4	41.6	17.0	46.0
30	40.7	45.4	20.2	42.8	17.8	47.2

Index

Statistical Methods in Practice: for Scientists and Technologists R. Boddy, G. Smith,
© 2009 John Wiley & Sons, Ltd